Recent Advances in Gas Chromatography

Edited by Fabrice Mutelet

Published in London, United Kingdom

IntechOpen

Supporting open minds since 2005

Recent Advances in Gas Chromatography
http://dx.doi.org/10.5772/intechopen.87748
Edited by Fabrice Mutelet

Contributors
Imadeddine Azzouz, Khaldoun Bachari, Andrew C. Bishop, Mohammad Sharif Khan, María Teresa García-Valverde, Carlos Ferreiro-Vera, Verónica Sánchez de Medina, Verónica Codesido, Jesús Hidalgo-García, Igor G. Zenkevich, Diana Margarita Hernandez-Baez, Alastair Reid, Antonin Chapoy, Bahman Tohidi, Roda Bounaceur, François Montel, Fabrice Mutelet, Amina Negadi, Amel Ayad

First published in London, United Kingdom, 2022 by IntechOpen
IntechOpen is the global imprint of INTECHOPEN LIMITED, registered in England and Wales, registration number: 11086078, 5 Princes Gate Court, London, SW7 2QJ, United Kingdom
Printed in Croatia

British Library Cataloguing-in-Publication Data
A catalogue record for this book is available from the British Library

Additional hard and PDF copies can be obtained from orders@intechopen.com

Recent Advances in Gas Chromatography
Edited by Fabrice Mutelet
p. cm.
Print ISBN 978-1-83962-602-9
Online ISBN 978-1-83962-603-6
eBook (PDF) ISBN 978-1-83962-613-5

For EU product safety concerns:
IN TECH d.o.o., Prolaz Marije Krucifikse Kozulić 3, 51000 Rijeka, Croatia, info@intechopen.com or visit our website at intechopen.com.

Meet the editor

Fabrice Mutelet is an associate professor in Chemical Engineering Thermodynamics at Ecole Nationale Superieure des Industries Chimiques – University of Lorraine (ENSIC-LRGP), France. He received his doctorate in 2001 from the University of Lorraine. He is a reviewer for more than twenty leading international journals and has published more than 120 research papers. His research interests include suitable sustainable solvents for chemical processes, the reduction of CO_2 emissions, and the measurement and correlation of phase diagrams for complex systems.

Contents

Preface

In the 1940s, Martin and Synge invented partition chromatography, laying the foundation for gas chromatography. Nowadays, the theory of chromatography and its basic principles are well described in the literature. Gas chromatography is probably one of the main techniques used in laboratories worldwide. Over the last several decades, researchers have pushed the limits of this technique by developing new procedures for the separation of components of systems found in biomedical or petrochemical industries. Inverse gas chromatography, a variation of conventional gas chromatography, was developed for the characterization of polymers, glass and carbon fibers, coal, and solid foods. In this technique, the material under investigation is placed in the chromatographic column and numerous probes are injected to provide information on the polarity or the surface of the sample.

This book includes contributions from experts in different domains. It beings with a chapter devoted to the identification and the validation of volatile organic compounds (VOCs) resulting from various diseases. It summarizes important technological advancements used to pre-concentrate and analyze VOCs. The next chapter describes recent advances in the analysis of Cannabis sativa L. by gas chromatography. The chapter includes two studies in which the thermal decomposition of analytes occurs during gas chromatography separation. Another chapter gives insight into the analysis of the reactive transport process occurring during the analysis of heavy oil hydrocarbons inside a high-temperature gas chromatography column. It also deals with those interrelated physicochemical processes generated by the application of a thermo-hydro-chemical (THC)-coupled multiphysics approach. The final chapter presents a thermodynamic model based on physico-chemical parameters measured using inverse gas chromatography.

This book is designed for those who have some acquaintance with gas chromatography, although we believe that it will be useful for beginners as well. Four chapters are devoted to specific techniques used in the medical and petrochemical industries.

Fabrice Mutelet
University of Lorraine,
Nancy, France

Chapter 1

Introductory Chapter: Recent Advances in Gas Chromatography

Fabrice Mutelet

1. Introduction

Gas chromatography (GC) is one of the most widely used techniques for the characterization, the separation and the quantification of complex systems. Researchers have pushed the limits of this technique by coming up with new methods for the preparation of samples and by using and/or coupling new families of columns. This last decade, the hyphenated technique coupling GC or GC-MS and a spectroscopic technique was developed [1]. These combinations of technologies have been used for qualitative but also quantitative studies of complex systems [2–5]. Multidimensional gas chromatography was also proposed for the analysis of complex fluids found in food, petroleum, and pharmaceutical industries [6–8]. It is now well established that comprehensive two-dimensional gas chromatography (GC × GC) is an efficient technique for fast pyrolysis bio-oil analysis [9], petroleum fluids [10], or characterization of flavonoid composition in food [11].

In gas chromatography, different approaches can be considered depending on the nature of the sample. Samples containing light compounds or moderate volatility can be studied using the classical approach. It means by the injection of the samples in the apparatus. For heavy compounds with low volatility, inverse gas chromatography (IGC) is preferred to characterize the samples. In IGC, the sample becomes the stationary phase. Both approaches do not give the same information, while IGC will give information on the interaction between a solute injected and the stationary phase or on the partition coefficient of the solute in the stationary phase, GC allows the quantification of components in the sample.

This last decade, new stationary phases based on ionic liquids or deep eutectic solvents were investigated due to their specific selectivity [12–16]. Numerous approaches were proposed to classify stationary phases [17]. Among others, Kovats index [18] and solvation models [19–22] are the most popular to represent the polar character of the stationary phases. All retention data related to Gibbs-free energy may be expressed using solvation models. Parameters from the linear solvation energy relationship (LSER) model can be estimated *via* gas chromatography. This approach was strongly used to develop relationship between physicochemical properties and LSER parameters [15, 23, 24].

In this book, state of the art of gas chromatography and new developments and applications are presented. New sample preparation techniques and hyphenated techniques are presented. The behavior and the characteristics of new stationary phases based on ionic liquids are also described. Then, theoretical approaches developed to predict the behavior of solutes with stationary phases are detailed.

IntechOpen

Conflict of interest

The authors declare no conflict of interest.

Author details

Fabrice Mutelet
Ecole Nationale Supérieure des Industries Chimiques, Université de Lorraine, Nancy, France

*Address all correspondence to: fabrice.mutelet@univ-lorraine.fr

References

[1] Patel KN, Patel JK, Patel MP, Rajput GC, Patel HA. Introduction to hyphenated techniques and their applications in pharmacy. Pharmaceutical Methods. 2010;**1**(1): 2-13. DOI: 10.4103/2229-4708.72222

[2] Wang FC-Y. Comprehensive two-dimensional gas chromatography hyphenated with a vacuum ultraviolet spectrometer to analyze diesel—A three-dimensional separation (GC × GC × VUV) approach. Energy Fuels. 2020;**34**(7):8012-8017. DOI: 10.1021/acs.energyfuels.0c00688

[3] Jia W, Fan Z, Du A, Li Y, Zhang R, Shi Q, et al. Recent advances in Baijiu analysis by chromatography based technology—A review. Food Chemistry. 2020;**324**:126899. DOI: 10.1016/j.foodchem.2020.126899

[4] Gordillo R. Supercritical fluid chromatography hyphenated to mass spectrometry for metabolomics applications. Journal of Separation Science. 2021;**44**(1):448-463. DOI: 10.1002/jssc.202000805

[5] Sdrigotti N, Collard M, Purcaro G. Evolution of hyphenated techniques for mineral oil analysis in food. Journal of Separation Science. 2021;**44**(1):464-482. DOI: 10.1002/jssc.202000901

[6] Machado ME. Comprehensive two-dimensional gas chromatography for the analysis of nitrogen-containing compounds in fossil fuels: A review. Talanta. 2019;**198**:263-276. DOI: 10.1016/j.talanta.2019.02.031

[7] Pico Y, Alfarhan AH, Barcelo D. How recent innovations in gas chromatography-mass spectrometry have improved pesticide residue determination: An alternative technique to be in your radar. TrAC Trends in Analytical Chemistry. 2020;**122**:115720. DOI: 10.1016/j.trac.2019.115720

[8] Kulsing C, Nolvachai Y, Marriott PJ. Concepts, selectivity options and experimental design approaches in multidimensional and comprehensive two-dimensional gas chromatography. TrAC Trends in Analytical Chemistry. 2020;**130**:115995. DOI: 10.1016/j.trac.2020.115995

[9] Negahdar L, Gonzalez-Quiroga A, Otyuskaya D, Toraman HE, Liu L, Jastrzebski JTBH, et al. Characterization and comparison of fast pyrolysis bio-oils from pinewood, rapeseed cake, and wheat straw using 13c nmr and comprehensive GC × GC. ACS Sustainable Chemistry & Engineering. 2016;**4**(9):4974-4985. DOI: 10.1021/acssuschemeng.6b01329

[10] Jennerwein MK, Sutherland AC, Eschner M, Gröger T, Wilharm T, Zimmermann R. Quantitative analysis of modern fuels derived from middle distillates—The impact of diverse compositions on standard methods evaluated by an offline hyphenation of HPLC-refractive index detection with GC×GC-TOFMS. Fuel. 2017;**187**:16-25. DOI: 10.1016/j.fuel.2016.09.033

[11] Cecchi L, Ieri F, Vignolini P, Mulinacci N, Romani A. Characterization of volatile and flavonoid composition of different cuts of dried onion (*Allium cepa* L.) by HS-SPME-GC-MS, HS-SPME-GC×GC-TOF and HPLC-DAD. Molecules. 2020;**25**(2):408. DOI: 10.3390/molecules25020408

[12] Nkosi N, Tumba K, Ramsuroop S. Activity coefficients at infinite dilution of various organic solutes in the deep eutectic solvent (tetramethylammonium chloride + 1,6 hexanediol in the 1:1 molar ratio). South African Journal of Chemical Engineering. 2019;**27**:7-15. DOI: 10.1016/j.sajce.2018.11.003

[13] Nkosi N, Tumba K, Ramsuroop S. Activity coefficients at infinite dilution

of various solutes in tetra propylammonium bromide + 1,6-hexanediol deep eutectic solvent. Journal of Chemical and Engineering Data. 2018;**63**(12):4502-4512. DOI: 10.1021/acs.jced.8b00600

[14] Nkosi N, Tumba K, Ramsuroop S. Measurements of activity coefficient at infinite dilution for organic solutes in tetramethylammonium chloride + ethylene glycol deep eutectic solvent using gas-liquid chromatography. Fluid Phase Equilibria. 2018;**462**:31-37. DOI: 10.1016/j.fluid.2018.01.019

[15] Mutelet F, Baker GA, Ravula S, Qian E, Wang L, Acree WE. Infinite dilution activity coefficients and gas-to-liquid partition coefficients of organic solutes dissolved in 1-sec-butyl-3-methylimidazolium bis(trifluoromethylsulfonyl)imide and in 1-tert-butyl-3-methylimidazolium bis(trifluoromethylsulfonyl)imide. Physics and Chemistry of Liquids. 2019;**57**(4):453-472. DOI: 10.1080/00319104.2018.1491045

[16] Revelli A-L, Mutelet F, Jaubert J-N. Prediction of partition coefficients of organic compounds in ionic liquids: Use of a linear solvation energy relationship with parameters calculated through a group contribution method. Industrial and Engineering Chemistry Research. 2010;**49**(8):3883-3892. DOI: 10.1021/ie901776z

[17] Abraham MH, Poole CF, Poole SK. Classification of stationary phases and other materials by gas chromatography. Journal of Chromatography A. 1999;**842**(1):79-114. DOI: 10.1016/S0021-9673(98)00930-3

[18] Kováts E. Gas-chromatographische Charakterisierung organischer Verbindungen. Teil 1: Retentionsindices aliphatischer Halogenide, Alkohole, Aldehyde und Ketone. Helvetica Chimica Acta. 1958;**41**(7):1915-1932. DOI: 10.1002/hlca.19580410703

[19] Abraham MH. Scales of solute hydrogen-bonding: Their construction and application to physicochemical and biochemical processes. Chemical Society Reviews. 1993;**22**(2):73-83. DOI: 10.1039/CS9932200073

[20] Abraham MH, Whiting GS, Doherty RM, Shuely WJ. Hydrogen bonding. XVI. A new solute salvation parameter, π2H, from gas chromatographic data. Journal of Chromatography A. 1991;**587**(2):213-228. DOI: 10.1016/0021-9673(91)85158-C

[21] Abraham MH, Whiting GS, Doherty RM, Shuely WJ. Hydrogen bonding. Part 13. A new method for the characterisation of GLC stationary phases - The Laffort data set. Journal of the Chemical Society, Perkin Transactions. 1990;**2**(8):1451-1460.

[22] Proctor A, Sprunger L, Acree WE Jr, Abraham MH. LFER correlations for the solubilising characterisation of room temperature ionic liquids containing trifluoromethanesulfonate and trifluoroacetate anions. Physics and Chemistry of Liquids. 2008;**46**(6):631-642. DOI: 10.1080/00319100802087191

[23] Mutelet F, Ortega-Villa V, Moïse J-C, Jaubert J-N, Acree WE. Prediction of partition coefficients of organic compounds in ionic liquids using a temperature-dependent linear solvation energy relationship with parameters calculated through a group contribution method. Journal of Chemical and Engineering Data. 2011;**56**(9):3598-3606. DOI: 10.1021/je200454d

[24] Rabhi F, Mutelet F, Sifaoui H, Wagle DV, Baker GA, Churchill B, et al. Characterization of the solubilizing ability of tetraalkylammonium ionic liquids containing a pendant alkyl chain bearing a basic N,N-dimethylamino or N,N-dimethylaminoethoxy functionality. Journal of Molecular Liquids. 2019;**283**:380-390. DOI: 10.1016/j.molliq.2019.03.066

Chapter 2

Reactive Transport and Its Implications on Heavy Oil HTGC Analysis – A Coupled Thermo-Hydro-Chemical (THC) Multiphysics Modelling Approach

Diana Margarita Hernandez-Baez, Alastair Reid, Antonin Chapoy, Bahman Tohidi, Roda Bounaceur and François Montel

Abstract

This chapter provides an insight into the reactive transport in a capillary column which heavy-oil hydrocarbons undergo when analysed by high temperature gas chromatography (HTGC), and their implications on characterisation outcomes, namely thermal cracking of the injected sample; and incomplete or non-elution of heavy components from the column, by using a coupled Thermo-Hydro-Chemical (THC) multiphysics modelling approach. For this purpose, a computational coupled THC, multicomponent, multi-physics model is developed, accounting for: multiphase equilibrium using an in-house, extended thermodynamics distribution factors dataset, up to $nC_{98}H_{198}$; transport and fluid flow in COMSOL and MATLAB; and chemical reactions using kinetics and mechanisms of the thermal cracking, in CHEMKIN. The determination of the former extended dataset is presented using two complementary HTGC modes: i) High-Efficiency mode, with a long column operated at low flow rate; and ii) true SimDist mode, with a short column operated at high flow rate and elution up to $nC_{100}H_{202}$.

Keywords: Reactive Transport, Thermo-Hydro-Chemical (THC) processes, coupled THC modelling, coupled multi-physics, multiphase equilibrium, solvation thermodynamics, transport and fluid flow, chemical reactions, kinetics and mechanism of thermal cracking, pyrolysis, Heavy n-alkanes thermodynamics distribution factors, High-temperature gas chromatography (HTGC), heavy-oil characterisation, Gas Chromatography modelling, coupled multiphysics modelling, CHEMKIN, COMSOL, MATLAB

1. Introduction

The objective of this chapter is to understand the reactive transport occurring during the High Temperature Gas Chromatography (HTGC) analysis of heavy oil hydrocarbons at common conditions and thereby quantify the implications on the final characterisation results in terms of: (i) the degree of thermal cracking of the original sample; and (ii) the non-elution of heavy components from the HTGC column by using a combined Thermo-Hydro-Chemical (THC) coupled multiphysics modelling approach.

For this endeavour, a synergy between experimental and computational coupled multi-physics approaches, has been carried out to account for three physicochemical processes: thermodynamic equilibrium fluid-flow; transport of chemical species; and chemical reactions.

An outline is given of the experimental approach used, with explanation of the methodology for extending the distribution factors data-set, necessary to describe the first process.

On the computational side, an in-house coupled multi-physics model has been developed using MATLAB [1] as language host, to couple the above three processes. The former is described, using as input to the multi-physics model the extended, experimental distribution factors dataset: and the latter two processes are described using: COMSOL Multi-physics [2] and MATLAB, and CHEMKIN [3] respectively.

Finally, the implication of the inter-related, multi-physics, physicochemical processes is discussed, based on the results from the coupled THC multi-physics model.

2. Experimental overview

2.1 Outline of HTGC methodology

Detailed accounts of the experimental procedures have been published previously [4], covering the generation of isothermal and temperature-programmed retention data for n-alkanes and polywax mixtures, on poly-dimethyl-siloxane HTGC columns, up to 430°C. (i.e. based on well-established SimDist techniques). This database then enabled the distribution factors of heavy n-alkanes to be derived up to $nC_{98}H_{198}$, which were unavailable in the literature.

2.2 Methodology for extending distribution factors up to $nC_{98}H_{198}$

In the absence of a database of distribution factors of heavy n-alkanes, it was necessary to obtain insight into their behaviour inside the HTGC column, requiring development of a comprehensive methodology to extend the existing, limited amount of data up to around $nC_{98}H_{196}$ [4].

Hernandez et al. [4] derived a temperature-dependent function of distribution factors which has been applied to a series of n-alkanes spanning ($nC_{12}H_{26}$ - $nC_{98}H_{196}$) by combining Eq. (1) and numerous isothermal experiments carried out using two *SGE* HT5 GC capillary columns [5] of different lengths, and under two HTGC methods as follows:

a. For the long column, low flow rates were used, to obtain efficient resolution of eluting n-alkanes, spanning the range ($nC_{12}H_{26}$–$nC_{64}H_{130}$), under constant inlet pressure measurements conditions.

b. For the short column, ASTM method D7169–11 [6] was applied to achieve extended SimDist analysis, spanning the range $nC_{12}H_{26}$–$nC_{98}H_{198}$ under constant flow rate measurement conditions.

In both columns, the standard samples (ASTM D5442) was used and at least 3 isothermal GC measurements were carried out from 80 to 420°C at 20°C intervals, and lastly at 430°C. Further details can be found in [4].

3. Theory

3.1 Physicochemical HTGC workflow

In order to understand the Reactive Transport (inter-related Thermo-Hydro-Chemical multi-physics processes) occurring during HTGC analysis of heavy oils, it is necessary to consider them step-wise, within the column.

- as the vaporised hydrocarbon/CS_2 solvent sample is transported through the column, the ***Diffusion Process*** of each component is considered to be negligible in comparison with the ***Convection Process*** [7].

- the stationary phase controls the adsorption of each component of the sample as a function of its boiling point in relation to the oven temperature program

- the vaporised hydrocarbon sample encounters the stationary phase, where each component experiences a ***gas–liquid thermodynamic equilibrium process*** at the prevailing temperature and pressure, between the stationary and mobile phases.

- Each component of the sample which has been dissolved in the stationary phase is retained at the same point until its vapour pressure equals that of the carrier gas, as the column temperature program proceeds, and is then transported to the GC outlet. It is assumed that ***no diffusivity effects occur in the liquid phase***.

- The solvent has a low boiling point compared to the hydrocarbon mixture, being the first component to be eluted or released by the stationary phase and transported by the carrier gas to the column outlet. It therefore generates the first peak in the gas chromatogram results, representing the least retained solute in the sample.

- Each component might encounter other components in the carrier gas and it is assumed that all of them are transported as a batch mixture.

- Each mixture is at risk of suffering a ***thermal cracking process*** at a given temperature in the gas phase, depending on whether the kinetics and mechanism at a given temperature reach the necessary energy to break some of the chemical bonds of the component, creating smaller chemical species.

- Each component of the sample whose boiling point is not reached during the temperature program, is retained by the stationary phase, and at risk of either incomplete elution, or non-elution from the column.

- The longer a heavy component is retained by the stationary phase before desorbing, the greater the risk of thermal cracking due to its greater exposure to the highest temperature.

3.2 Thermo-hydro-chemical processes

Thus, the three physicochemical processes in a heavy-oil HTGC analysis considered in this work are:

- Multiphase equilibrium:

  Solvation thermodynamics using experimental data.

- Transport and fluid flow:

  Convection in MATLAB

- Chemical reactions of thermal cracking:

Kinetics and mechanisms of simplified mixture in CHEMKIN,
which fall within the classification *Thermo-Hydro-Chemical multi-physics* [8].

3.2.1 Multiphase equilibrium: solvation thermodynamics

The basis of the gas chromatography separation process centres on the non-isothermal multiphase equilibrium of each of the chemical species in the mixture sample between the stationary phase and the gas phase (transported by the carrier gas) taking advantage of their distinct boiling points.

This equilibrium is established in multiple stages throughout the length of the capillary column. The analysis mixture sample is dissolved and retained in the stationary phase at low initial temperatures and each component comprising the mixture is evaporated and separated from the sample mixture once its boiling point temperatures and pressure is reached. Thus, solvation thermodynamics is used to describe the gas–liquid equilibrium of each chemical species inside the GC column.

The temperature-dependent expression for the distribution factor, K, is described in Eq. (1) and was obtained by solving the thermodynamic equilibrium of the solvation of a solute in the bulk solvent [9] expressed in terms of the Gibbs free energy at a given temperature and by the logarithm of the solute molecule's numeral density ratio in both phases [10, 11] or the ratio between the molar concentration of the two phases.

The mass transfer is assumed to be governed only by the interaction between the solute and the stationary phase, while the interactions between solute-solute and solute-carrier gas are assumed to be negligible as the interfacial and extra-column effects leading to non-equilibrium conditions [12].

A semi-empirical model [13, 14] developed by Castells et al. [15] for the determination of the isothermal retention times as function of the hold-up time, t_M and the solvation time expressed by the Gibbs free energy is expressed in the terms of ΔH and ΔS, which represent the changes in enthalpy and entropy associated with the transfer of solute from the stationary phase to the mobile phase at a given temperature T.

$$K(T(t)) = \beta\left[\frac{t_r}{t_M} - 1\right] = \exp\left[a_0 + a_1\frac{1}{T(t)}\right]$$

$$a_0 = \frac{\Delta S(T)}{R}; a_1 = -\left(\frac{\Delta H(T)}{R}\right)$$

$$\beta = \frac{(2r_o - 2w)^2}{2{r_o}^2 - (2r_o - 2w)^2} \tag{1}$$

In Eq. (1), K corresponds to the distribution factor, with t_r and t_m representing the retention time of the solute and the hold-up time, respectively. β is the phase ratio of the column, r_o and w correspond to the inner radius of the column, and the film thickness of the stationary phase. ΔH and ΔS, correspond to the delta changes in enthalpy and entropy associated with the transfer of solute from the stationary phase to the mobile phase.

Aldaeus [16] has proposed two retention mechanisms according to the nature of the separation hold between the analyte and the stationary phase, based on the semi-empirical values of the thermodynamic properties of Eq. (1).

3.2.2 Transport and fluid flow: convection

The Snijders [17] method for calculating the retention times in gas chromatography is based on the peak position determination which is not affected by the diffusion effects but by the convection process only [16].

Eq. (2) expresses the convection process in terms of the effective velocity v_{eff} of the analyte in the carrier gas. Discretized into finite time-steps of Eq. (2) allows tracking of the position of the analyte at every x position through the GC column at every time step until the peak reaches the column outlet [18, 19] at the final time step which cumulated represents the retention time for that analyte as explained in [4]. And lastly, K and β are the distribution factor and phase ratio of the column described in Eq. (1) and v_m is the velocity of the mobile phase.

$$v_{eff,i}(x,t) = \frac{v_M(x,t)}{1 + \frac{K_i(T)}{\beta}} \tag{2}$$

Integrating the differential form of the Hagen-Poiseuille fluid mechanics Equations [10, 18] through the length of the column allows calculation of v_m as described in Eq. (3). This expression relates the carrier gas velocity to the pressure gradient at any position in the column [18] by a proportional constant. The latter depends of the geometry of the cross-section which in this case is for a column of circular cross-section [20]:

$$v_M(x,t) = \frac{{r_o}^2 \cdot \left({P_{in}}^2 - {P_{out}}^2\right)}{16 \cdot \eta(T(t)) \cdot L \cdot P(x)} \tag{3}$$

In Eq. (3), $\eta(T(t))$ corresponds to the viscosity of the carrier gas [21, 22]. (See summarised details in [19]), P_{in} and P_{out} are the inlet and outlet pressures of the GC column. r_o is the inner radius of the column and $P(x)$ is the pressure at position x

	Pyrolysis mechanisms	
	Radicals	**"Class" Molecular**
Mixture of Heavy Oils	nC_{14}, nC_{16}, nC_{20}, nC_{25}, nC_{30}, nC_{35}, nC_{40}, nC_{45}, nC_{50}, nC_{55}, nC_{60}, nC_{65}, nC_{70}, nC_{75}, nC_{80}	nC_{14}, nC_{16}, nC_{20}, nC_{25}, nC_{30}, nC_{35}, nC_{40}, nC_{50}, nC_{60}, nC_{70}, nC_{80}
Reactants	15	11
Reactions	7055	127
Species	336	17
Molecules	242	
Radicals	94	

Table 1.
Summary of size of the mechanistic kinetics models developed.

described with Eq. (4), which is obtained by integrating the Hagen-Poiseuille equation between the inlet and outlet position, of a differential element and assuming incompressibility of the gas in each element at position x, due to the extremely low pressure-drop in gas chromatography [10].

$$P(x) = \sqrt[2]{{P_{in}}^2 - \left({P_{in}}^2 - {P_{out}}^2\right)\frac{x}{L}} \tag{4}$$

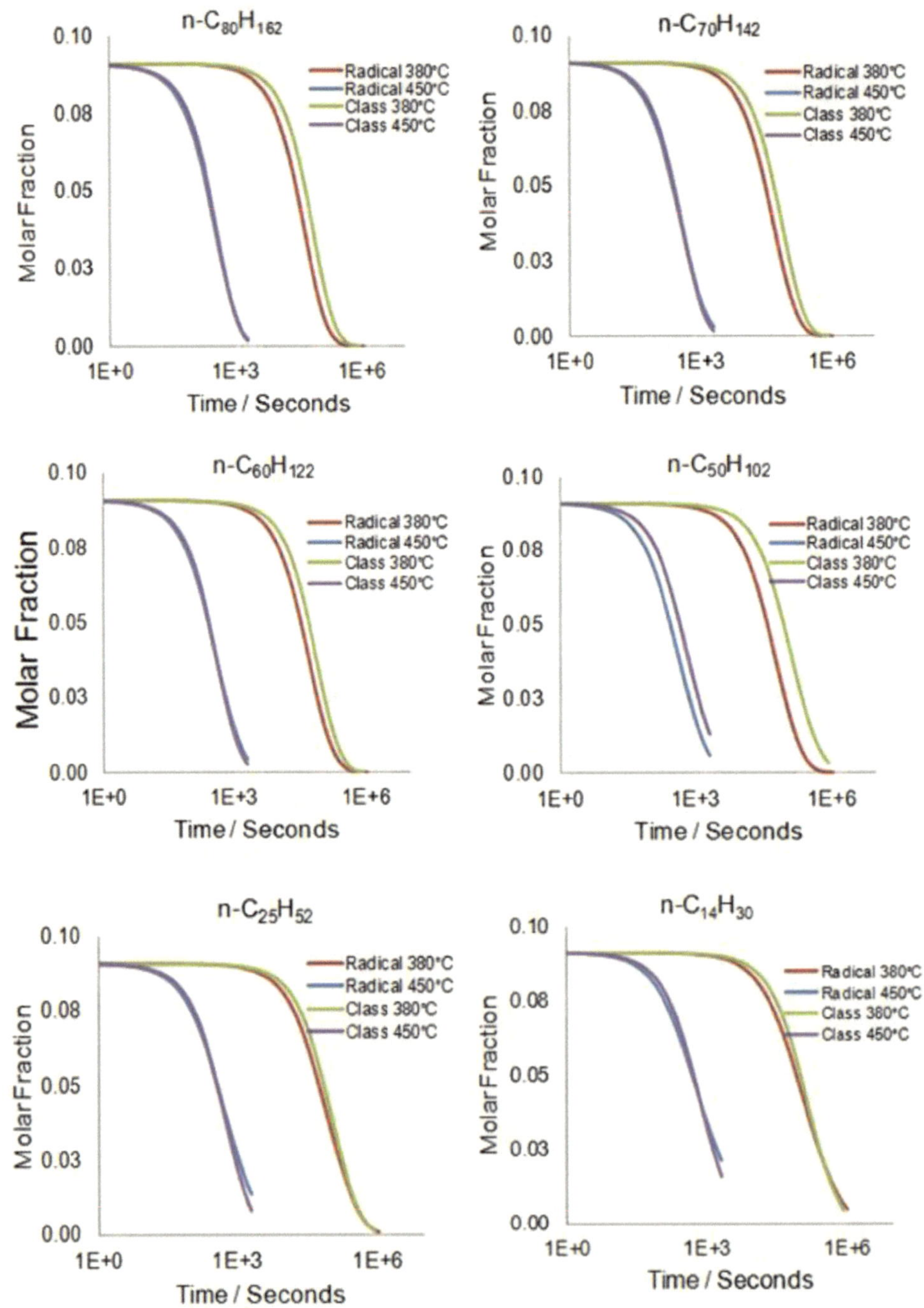

Figure 1.
Comparison of free radical model and "class" molecular model for heavy n-alkanes mixtures. (simulation of a closed reactor at 1 MPa).

3.2.3 Chemical reactions: kinetics and mechanism of thermal cracking

The large number of species in the reduced free-radical pyrolysis model developed in [23] has imposed a need to develop a reduced molecular pyrolysis model, comprising 11 n-alkanes ($nC_{14}H_{30}$, $nC_{16}H_{32}$, $nC_{20}H_{42}$, $nC_{25}H_{52}$, $nC_{30}H_{62}$, $nC_{35}H_{72}$, $nC_{40}H_{82}$, $nC_{50}H_{102}$, $nC_{60}H_{122}$, $nC_{70}H_{142}$, and $nC_{80}H_{162}$).

In this work, a "class" molecular mechanism has been obtained after applying the following three rearrangements to our reduced molecular mechanism model [7]:

- Lumping of molecules belonging to the global class "C_{15} plus" which are produced by an n-alkane reactant.
- Lumping of n-alkane reactants, which produced n-alkane reactants or lighter class.
- Lumping of global class C_{15} as reactant.

Refer to [23], to understand the thermal cracking original kinetic and mechanism model development, and refer to [7] to understand the detailed explanation of the kinetics and mechanism reduction procedure from molecular mechanism model to a "class" molecular mechanism.

The optimised reduced "class" molecular mechanism used in this work is composed of 127 molecular reactions and 17 species (11 n-alkanes, and 6 "class" molecular pyrolysis products) which has been obtained after applying to the whole mechanism the above rearrangement and its corresponding kinetic data [7].

Thus, the final reduced molecular mechanism, accounts for:

- 11 original n-alkanes (reactants): nC_{14}, nC_{16}, nC_{20}, nC_{25}, nC_{30}, nC_{35}, nC_{40}, nC_{50}, nC_{60}, nC_{70}, nC_{80}.
- 6 classes: alkene, CH_4, C_2H_6, C_3-C_5, C_6-C_{13} and C_{15} plus.

Finally, as summarised in **Table 1** the number of reactions of the original free-radical pyrolysis model has been reduced from 7055 to 127, and the number of species from 336 to 17, whilst still yielding very good accuracy as depicted in **Figure 1**.

4. Computational multi-physics

An understanding of the Thermo-Hydro-Chemical (THC) processes occurring inside an HTGC column during the analysis of heavy oil hydrocarbons was obtained through detailed study with an in-house coupled THC model.

The coupling of the physico-chemical processes is sequential due to the complexity of the system, and the level of detail with which each process has been described. Hence, a fully coupled model is prohibited while a sequential coupling can handle effectively the following processes:

- the thermodynamics equilibrium is described using experimental input data of a series of n-alkanes spanning ($nC_{12}H_{26}$-$nC_{98}H_{196}$) [4].
- the chemical reactions are described using kinetics and mechanistic modelling of 11 n-alkanes: nC_{14}, nC_{16}, nC_{20}, nC_{25}, nC_{30}, nC_{35}, nC_{40}, nC_{50}, nC_{60}, nC_{70}, nC_{80} and 6 class molecules: alkene, CH_4, C_2H_6, C_3-C_5, C_6-C_{13} and C_{15} plus.

- the convection process is described using the Hagen-Poiseuille fluid mechanics equations [10, 18].

The sequence of these processes was arranged using short time intervals where the temperature was constant during the temperature ramp, and with a batch size as described using Golay's theory [24] for diffusion and convection processes.

4.1 Computational HTGC workflow

From a computational modelling perspective, the multi-physics processes can be simplified and described as follows:

1. The system to model is delimited to the gas phase inside the HTGC column, comprising only the carrier gas transporting each component from the column inlet to the outlet.

2. The control volume is the inner volume of the coiled capillary GC column (e.g.,typically 0.53 mm internal diameter, and 25-30 m length).

3. The fluid flow and transport of chemical species considers only the convection process of the carrier gas transporting each of the species at its own speed from the GC inlet to the GC outlet. The diffusion of the chemical species in the gas phase has been concluded to be negligible based on a previous investigation [19] which compares the advection process (convection plus diffusion) and the convection process only, demonstrating no difference between the chromatogram peaks results using both approaches.

4. The equilibrium occurring in the interface between the stationary phase and the gas phase is modelled through the extended thermodynamics distribution factors dataset obtained previously [4], for each one of the heavy n-alkanes hydrocarbons mixture.

5. The heavy oil mixture is simplified to one comprising only heavy n-alkanes, ranging from $nC_{25}H_{52}$ to $nC_{98}H_{198}$ which is suitably representative, bearing in mind that long-chain n-alkanes are the most susceptible to thermal cracking.

6. Their thermal cracking is modelled in the gas phase only, using a reduced and optimised kinetics and mechanism developed previously [7] and validated against a detailed mechanism of the heavy oil mixture developed initially [23].

4.2 Coupled multi-physics workflow

Finally, the coupling of the three physics involved is made in a sequential order as follows:

a. The column is treated as a series of batches of 1 mm

b. The time for the multi-physics processes to occur in each batch is the time for each species to travel at the carrier gas velocity, from the centroid of one batch to the following one, with velocity derived from volumetric flow measurement and column i.d.

c. The chemical species are modelled through convection only, using MATLAB. No diffusivity is included in this final work as explained in [7]

d. The chemical species undergo an equilibrium which is reached based on the thermodynamics of solvation using the extended distribution factors data set introduced previously [4]

e. The position of each of the chemical species is calculated and each batch is created according to the chemical species found in the batch volume. Thus, the chemical reactions of the thermal cracking are modelled in each mini-batch reactor.

f. Each batch is updated with the new species found inside.

g. The convection of the new chemical species is modelled as described in c) and so on, until all the batches arrive at the GC outlet sequentially.

4.3 Discretization methods

This work uses the discretization method introduced by Snijders [17], which predicts the peak width of the solute zone as the space that a solute migrating through the column occupies [25]. This approach of the convection model was successfully coupled to the reduced molecular pyrolysis model from [7].

Equal time segments are used to discretize the simulation as proposed by Snijders [17] for enabling isothermal properties to be used at every time-step according to the ramp of temperature used. Thus, a sufficiently small time-step permits a uniform pressure to be assumed in the column segments traversed by the solute.

The local plate height (H) is calculated at every time-step based on the Golay [24] expression for open tubular columns, as shown in Eq. (5), where k is the retention factor.

$$H(x,t) = 2\cdot\frac{D_M(x,t)}{v_M(x,t)} + v_M(x,t)\left\{\left[\frac{1+6\cdot k(T(t))+11\cdot k(T(t))^2}{24\cdot[1+k(T(t))]^2}\cdot\frac{r_o{}^2}{D_M(x,t)}\right] + \left[\frac{2\cdot k(T(t))}{3\cdot[1+k(T(t))]^2}\frac{w^2}{D_s(x,t)}\right]\right\} \tag{5}$$

Note here that k is dimensionless, being derived from the distribution factor, K, and the phase ratio of the column, β namely K/β, with K corresponding to the ratio between the (moles/volume) in stationary phase to the (moles/volume) in gas phase. r_o and w correspond to the inner radius of the column and the film thickness of the stationary phase D_s, and D_m correspond to the diffusion constant respectively in the stationary and mobile phase. v_m corresponds to the velocity of migration of the carrier gas.

The local zone variance (σx^2) in the distance of a solute from the zone centroid at a given position x, can be calculated using Eq. (6), representing the solute band's spreading.

$$\sigma_x{}^2(\Delta x_n) = H(x_n, t_n)\cdot\Delta x_n \tag{6}$$

Eq. (7) describes the increment in the zone variance length, and can be obtained by the summation of all the local contributions of zone variances. Giddings [26] explained that at every time step, the correction is applied for the expansion of the solute zone due to the reduction in pressure (P) along the column.

$$\sigma_x{}^2(x_n) = \left[\sum_{i=1}^{n-1} \sigma_x{}^2(\Delta x_i)\right] \cdot \left\{\frac{P(x_{n-1})}{P(x_n)}\right\} + \sigma_x{}^2(\Delta x_n) \tag{7}$$

This approach, has been programmed in MATLAB, and has been compared in [7] with the solution yielded by the COMSOL-MATLAB model developed in [19], which solves the diffusive-convective equation by finite elements.

The comparison study confirmed excellent agreement in predictions of the zone's centroid (average relative error of 1.1%) and of the zone's standard deviations (average relative error of 3%), as depicted in **Figure 2**.

Thus, the analytical model implemented in MATLAB has been coupled to the reduced molecular pyrolysis model described above, and as detailed in [7], by calling CHEMKIN at every time step iteration, and using feedback between the two models until each component elutes from the GC column.

4.4 Coupled thermo-hydro-chemical processes

Both the reduced molecular pyrolysis model and the analytic iterative GC model derive from the prior high-performance improvement process required for ultimately attaining an efficient coupled physics-chemical model.

The latter can predict the zone's centroid, the standard deviation and the pyrolysis decomposition of every solute studied for both as a mixture and as a single component according to the position of every solute related to the batch width at every time-step.

In order to maintain a constant temperature at every time-step, a constant time-step has been implemented, permitting an increment of 1°C every 4 seconds (corresponding to the ramp of 15°C/min, used).

Initially, for every component studied, the position of the zone's centroid in the next time step (x_{i+1}), is calculated, using Snijders [23, 27] approach Eq. (8) (see ref. [19]), the distribution factor (K), and the phase ratio (β).

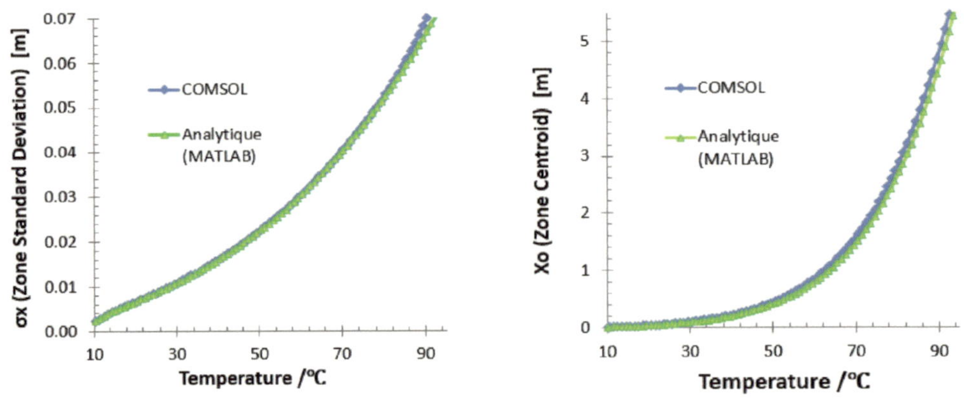

Figure 2.
*Comparison of zone standard deviation and zone centroid of $nC_{12}H_{26}$, predicted using an iterative analytic approach[11] using MATLAB and solving the diffusive-convection equation by finite element in COMSOL. (Column dimensions **Table 3** and temperature programming **Table 2**).*

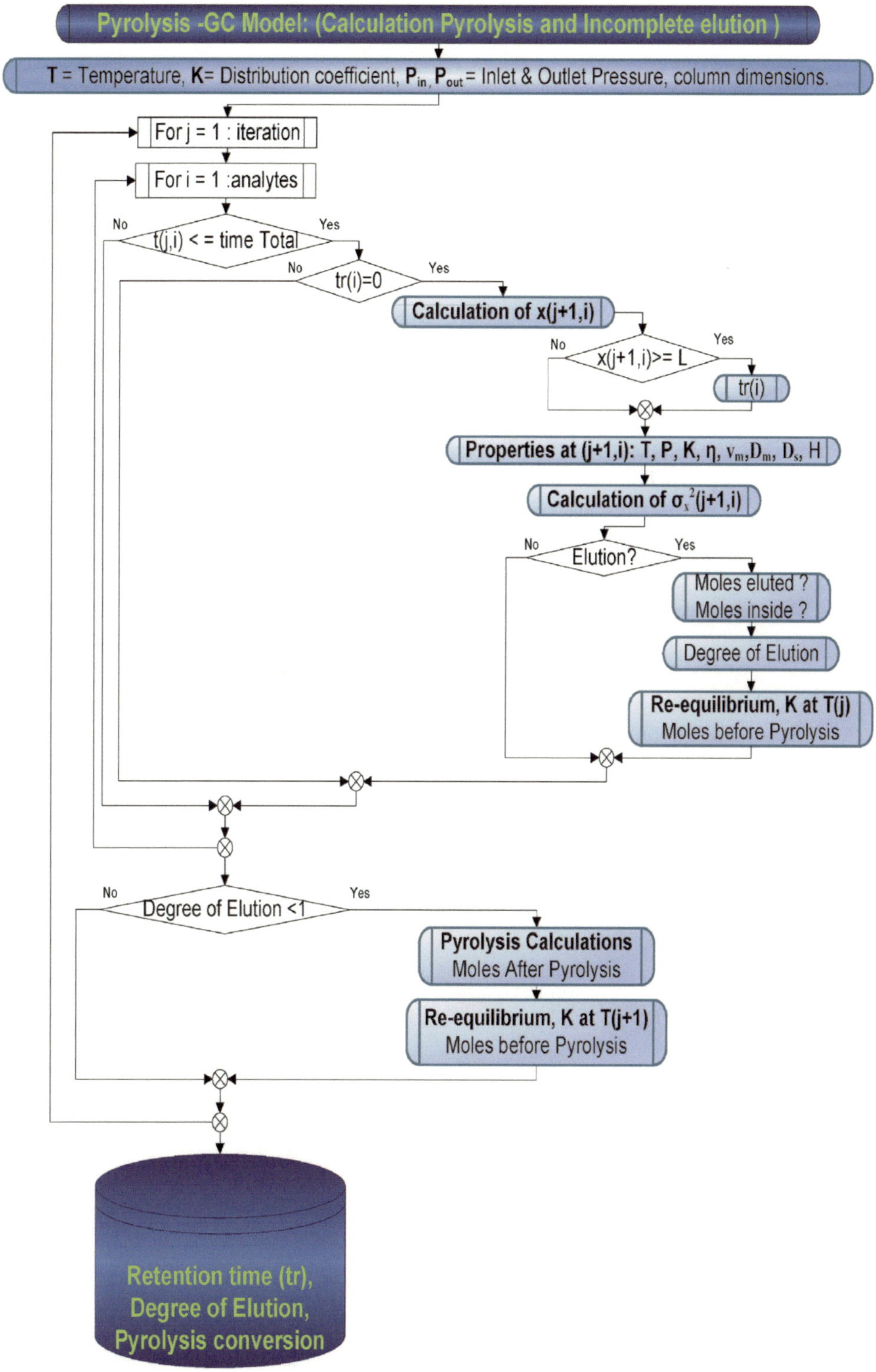

Figure 3.
Algorithm of the pyrolysis-GC coupled model.

$$x_{i+1} = x_i + \frac{\upsilon_M(x_i, t_i)}{1 + \frac{K_i(T(t_i))}{\beta}} \cdot \Delta t \tag{8}$$

Figure 3, shows the algorithm explaining the global calculation carried out by the coupled THC model for an heavy oil analysis by HTGC, using the above models as explained previously. For more detail refer to [7].

5. Implications: results

The implications of the THC processes during an HTGC heavy oil analysis can be summarised under two headings:

Thermal cracking risk of the original sample.

Non-elution or incomplete elution of the sample.

A detailed analysis of these implications is presented, based on the results of the in-house developed THC multiphysics model, described above.

5.1 Thermal cracking risk of the original sample

The cumulative conversion due to pyrolysis of the 11 n-alkanes is studied in [4], in order to analyse their risk, as depicted in **Figure 4**. For each component the ratio is calculated of the cumulative mass lost (**Figure 5**) due to thermal cracking, compared to the mass injected.

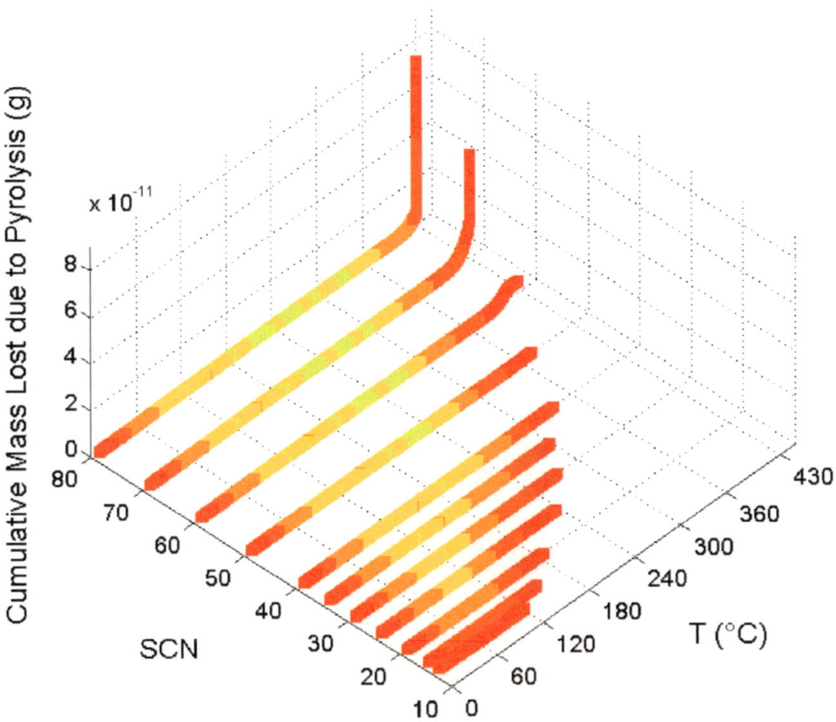

Figure 4.
*Accumulative mass lost due to thermal cracking for n-alkanes (nC_{14}, nC_{16}, nC_{20}, nC_{25}, nC_{30}, nC_{35}, nC_{40}, nC_{50}, nC_{60}, nC_{70}, nC_{80}) at a common HTGC temperature programming (**Table 2**) in a HT5 column with dimension summarised in (**Table 3**).*

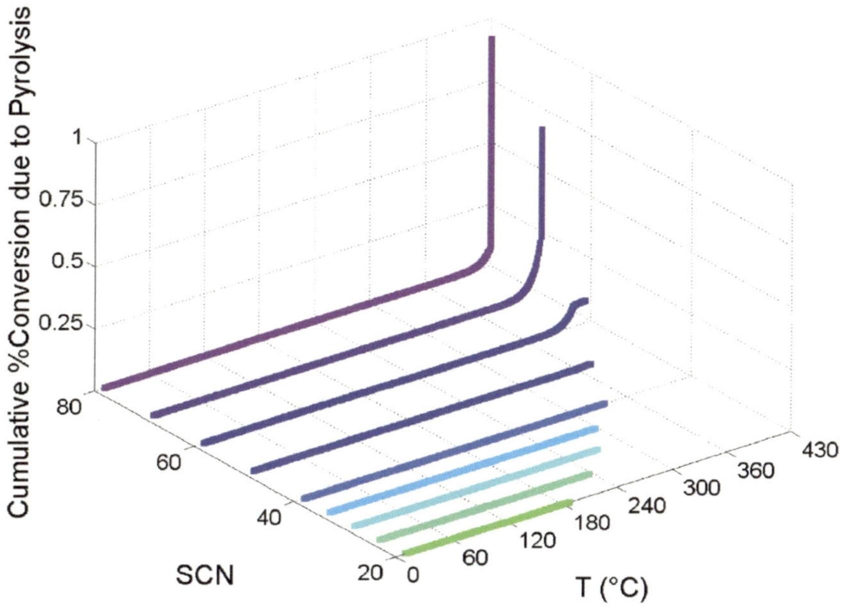

Figure 5.
*Cumulative conversion due to thermal cracking for n-alkanes (nC_{14}, nC_{16}, nC_{20}, nC_{25}, nC_{30}, nC_{35}, nC_{40}, nC_{50}, nC_{60}, nC_{70}, nC_{80}) at a common HTGC temperature programming (**Table 2**) in a HT5 column with dimension summarised in (**Table 3**).*

Parameters	**Values**
$T_{initial}$ (°C)	10
Hold up time at Tinicial (min)	0
ramp of T (°C/min)	15
T_{max} (°C)	430
Hold up time at T_{max} (min)	12

Table 2.
Temperature programming.

As would be expected, no pyrolysis reaction occurs in the case of $nC_{14}H_{30}$ and $nC_{16}H_{34}$ with the temperature program used (**Table 2**), and their associated short residence times inside the GC column. Similarly, within the range $nC_{20}H_{42}$ to $nC_{40}H_{82}$, insignificant conversion occurs, whereas in the case of $nC_{50}H_{102}$ the maximum mass loss through thermal decomposition before elution is 0.003%.

Low but detectable mass loss occurs with the heaviest n-alkanes. $nC_{60}H_{122}$ has a significant loss in the stationary phase where only $2.43 \cdot 10^{-12}$ g are released to the gas phase with the remainder trapped in the stationary phase. Further, pyrolysis loss begins at 373°C with a 0.001% cumulative mass conversion. $nC_{70}H_{142}$, presents a cumulative conversion of 0.001% at 385°C with only $2.32 \cdot 10^{-10}$ g released in the gas phase and the rest trapped in the stationary phase.

It should be noted that at the time-step when $nC_{60}H_{122}$ decomposition starts, $nC_{50}H_{102}$ is virtually totally eluted (99.9%) eluted, and hence the pyrolysis products present no risk of co-elution with the latter. Rather, the pyrolysis products of $nC_{60}H_{122}$ are released gradually, evidenced by a slowly increasing baseline signal.

Similarly, $nC_{70}H_{142}$ starts to decompose when located 1.02 m away from the GC inlet, and 0.68 minutes after $nC_{60}H_{122}$ is essentially fully eluted (99.99%).

Therefore, the pyrolysis products present no risk of co-elution with, nor distortion of the peak for $nC_{60}H_{122}$.

Lastly, $nC_{80}H_{162}$ starts to decompose at 0.41 m from the column inlet, while $nC_{70}H_{142}$ is located 1.64 m from the inlet. Thus, when $nC_{70}H_{142}$ is essentially fully eluted (99.99%), at 7.83 m from the column inlet, $nC_{80}H_{162}$ has undergone a cumulative conversion of 0.52% mass loss by pyrolysis, relative to mass injected. That equates to $3.97.10^{-11}$ g of $nC_{80}H_{162}$ converted into pyrolysis products, and which co-elutes with $nC_{70}H_{142}$, resulting in unreliable quantification.

5.2 Non-elution of heavy components from the column

For the determination of non/incomplete elution of heavy n-alkanes, the data set of distribution factors of the n-alkanes spanning the range from $nC_{12}H_{26}$ to $nC_{98}H_{198}$, [4] was used as main input for the calculation of the degree of elution of each of the n-alkanes studied.

The *degree of elution* has been introduced in order to determine the non/incomplete elution of heavy n-alkanes (as explained in [19]) as depicted in **Figure 6**.

Alkanes heavier than $nC_{60}H_{122}$ elute during the isothermal plateau of the temperature programmed at 430°C. Therefore, constant distribution factors apply for the re-equilibration period, when characteristic peak broadening is observable. (c.f. the essentially symmetrical peaks associated with temperature programmed GC analyses).

	SGE HT5 GC Column
Length [m]	12
Diameter [mm]	0.53
Film thickness [um]	0.15

Table 3.
Column dimensions of in-house HTGC.

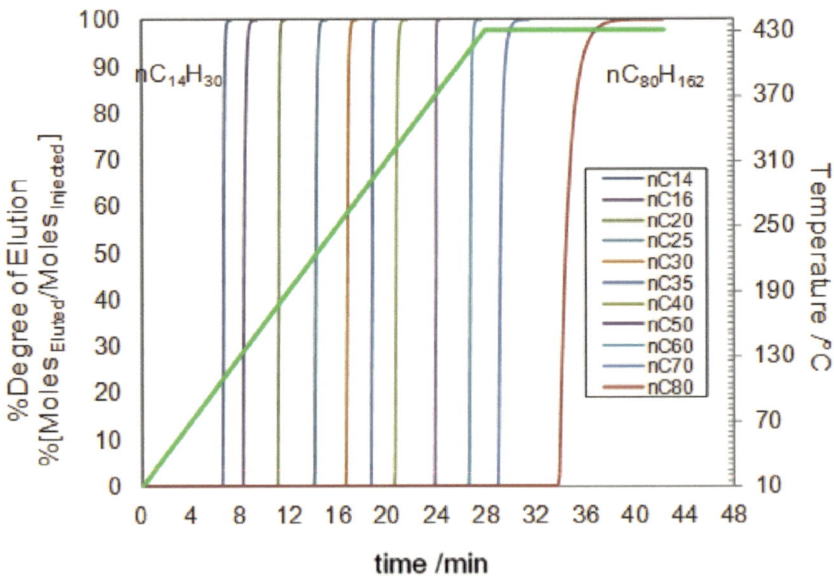

Figure 6.
Degree of elution vs. transit time of each component "i": n-alkanes in the range of $C_{14}H_{30}$ to $nC_{80}H_{162}$. Degree of elution = moles of "i" inside the GC column at time (t) /moles injected of "i".

$nC_{70}H_{142}$ starts to elute at 29 minutes with a 99.99% degree of elution at 31.3 minutes, and 100% at 31.5 minutes. $nC_{70}H_{142}$ takes 2.5 minutes to elute completely.

$nC_{80}H_{162}$ starts to elute at 33.8 minutes, with a degree of elution of 99.99% at 40.9 minutes and 100% after 42.3 minutes. $nC_{80}H_{162}$ takes 7.1 minutes to elute and 8.5 minutes to completely elute.

In this simulated study, components from $nC_{70}H_{142}$ and above, elute so slowly that peak resolution for the group cannot be assessed. Rather, in practice, a continuum is observed, in the form of a gradually increasing baseline, rising to a plateau which gradually reduces during the final isothermal period of the oven temperature program.

It is interesting to note that 99.99% of $nC_{80}H_{162}$ requires to elute 12.9 minutes at the isothermal conditions at the maximum temperature (430°C) of the analysis. of 99.99%. This means that this component is not normally taken into account in the GC calculations, due to the shorter period of time and stationary phase bleeding.

6. Conclusions

This chapter provides an insight into the analysis of the Reactive Transport process occurring during the analysis of heavy oil hydrocarbons inside a High Temperature Gas Chromatography column, and the implication that those interrelated physicochemical processes generate, by application of a Thermo-Hydro-Chemical (THC) coupled multiphysics approach.

The number of species in the reduced free-radical pyrolysis model developed in [19] has imposed a need to develop a reduced molecular pyrolysis model, comprising 11 n-alkanes ($nC_{14}H_{30}$, $nC_{16}H_{32}$, $nC_{20}H_{42}$, $nC_{25}H_{52}$, $nC_{30}H_{62}$, $nC_{35}H_{72}$, $nC_{40}H_{82}$, $nC_{50}H_{102}$, $nC_{60}H_{122}$, $nC_{70}H_{142}$, and $nC_{80}H_{162}$). The number of reactions has been reduced from 7055 to 127, and the number of species from 336 to 17, whilst still yielding very good accuracy.

THC multi-physics model has been implemented to resolve the HTGC limitations. The cumulative pyrolysis conversion of the 11 n-alkanes studied in this work, suggests that 0.52% of the mass injected of $nC_{80}H_{162}$, thermally decomposed before $nC_{70}H_{142}$. Therefore, co-elution of $nC_{70}H_{142}$ and the pyrolysis product of $nC_{80}H_{162}$ makes the GC analysis of $nC_{70}H_{142}$ and heavier n-alkanes no longer reliable.

The degree of elution of the 11 n-alkanes studied in the chapter confirms that alkanes heavier than $nC_{70}H_{142}$ take progressively longer to elute completely from the column, viz. $nC_{70}H_{142}$ takes 2.3 minutes and $nC_{80}H_{162}$ takes 7.1 minutes, with co-elution of decomposition products in each case compromising their analyses.

Finally, $nC_{80}H_{162}$ takes 12.9 minutes to completely elute during the isothermal plateau, resulting in no distinct peak being observable. Consequently, the eluting component will be masked in the FID plateau signal, in combination with column bleed products. As a result the $nC_{80}H_{162}$ analysis may not be utilised under these HTGC conditions.

Acknowledgements

The authors wish to thank the members of our JIP: Marathon Oil Corporation, Schlumberger and Total for both their technical and financial support during this project.

List of symbols

D_{eff}	Effective average Diffusivity (unit length^2/unit time)
D	Apparent diffusion coefficient, represents all factors causing dispersion (unit length^2/unit time)
D_M	Diffusion constant, mobile phase (unit length^2/unit time)
D_S	Diffusion constant, stationary phase (unit length^2/unit time)
H(x,t))	Column plate height, spatial rate of dispersion of a zone (unit length)
K	Distribution factor of a compound (moles/volume) in stationary phase/(moles/volume) in gas phase)
k	retention factor of a compound (moles in stationary phase/moles in gas phase).
L	Length of the GC column (unit length)
m(x,to)	mass profile for every analyte (particles/unit length).
$N_{i,M}$	Moles of component "*i*" in the mobile phase.
$N_{i,S}$	Moles of component "*i*" in the stationary phase.
P(x)	Pressure at position *x* (Pa)
Pin	Pressure at the GC colum inlet.
Pout	Pressure at the GC column outlet (Pa)
r_o	Internal radius of GC column (unit length)
ramp T	Ramp of temperature of the temperature programmed.
T(t)	Temperature at the time *t*.
To	Initial temperature of the temperature programmed.
t	time (unit time)
v_{eff}	Effective cross-sectional average velocity (unit length/unit time)
v_M	Velocity of migration of the carrier gas (unit length/unit time)
w	Film thickness (unit length)
X_i	Fraction of component *i* in the gas phase relative to the moles in both stationary and gas phase.
x_0	Centroid of Gaussian distribution of distribution of component inside the GC column (unit length)
x	Position of the component's dispersal around the centroid x_0. (unit length)

Greek letters

σ	Standard deviation of the distribution of component inside the GC column (unit length)
β	Phase ratio (volume of mobile phase in the column to the volume of stationary phase).
η_m	Viscosity of the carrier gas.(μPa·s).
Δt	Time step (unit time).

Author details

Diana Margarita Hernandez-Baez[1*], Alastair Reid[2†], Antonin Chapoy[2], Bahman Tohidi[2], Roda Bounaceur[3] and François Montel[4,5†]

1 SIMULEX Limited, International House, Edinburgh, Scotland, United Kingdom

2 Hydrates, Flow Assurance and Phase Equilibria Research Group, Institute of GeoEnergy Engineering, Heriot-Watt University, Edinburgh, Scotland, United Kingdom

3 Laboratory of Reactions and Process Engineering, Université de Lorraine, LRGP site ENSIC, NANCY Cedex, France

4 Fluids and Organic Geochemistry Department, Fluids Thermodynamics, Exploration and Production, Geosciences Technologies, TOTAL S.A, Avenue Larribau, France

5 Laboratoire des Fluides Complexes et leurs Réservoirs, LFCR, UMR 5150, Université de Pau et des Pays de l'Adour, E2S UPPA, CNRS, TOTAL, Pau, France

*Address all correspondence to: dianahernandez@gmail.com

† Retired.

References

[1] MATLAB. version 7.10.0 (R2010a). Natick, Massachusetts: The MathWorks Inc.; 2010.

[2] Multiphysics C. Introduction to COMSOL multiphysics extregistered. COMSOL Multiphysics, Burlington, MA, accessed Feb. 1998;9:2018.

[3] CHEMKIN 10112, Reaction Design: San Diego, 2011.

[4] Hernandez-Baez DM, Reid A, Chapoy A, Tohidi B. Determination of distribution factors for heavy n-alkanes (nC12-nC98) in high temperature gas chromatography. J Chromatogr A. 2019; 1591:138–146.

[5] SGE. HT5 GC Columns. http://www.sge.com/products/columns/gc-columns/ht5.

[6] ASTM D7169-11. (Standard Test Method for boiling Point Distribution of Samples with Residues such as Crude Oils and Atmospheric and Vacuum Residues by High Temperature Gas Chromatography).

[7] Hernandez-Baez DM, Reid A, Chapoy A, Tohidi B, Bounaceur R. Establishing the Maximum Carbon Number for Reliable Quantitative Gas Chromatographic Analysis of Heavy Ends Hydrocarbons. Part 3. Coupled Pyrolysis-GC Modeling. Energy Fuels. 2019 Mar 21;33(3):2045–2056.

[8] Keyes DE, Mcinnes LC, Woodward C, Gropp W, Myra E, Pernice M, Bell J, Brown J, Clo A, Connors J, Constantinescu E, Estep D, Evans K, Farhat C, Hakim A, Hammond G, Hansen G, Hill J, Isaac T, Jiao X, Jordan K, Kaushik D, Kaxiras E, Koniges A, Lee K, Lott A, Lu Q, Magerlein J, Maxwell R, Mccourt M, Mehl M, Pawlowski R, Randles AP, Reynolds D, Rivière B, Rüde U, Scheibe T, Shadid J, Sheehan B, Shephard M, Siegel A, Smith B, Tang X, Wilson C, Wohlmuth B. Multiphysics simulations. Int J High Perform Comput Appl. 2013;27(1):4–83.

[9] Giddings JC. Dynamics of chromatography, Par I, Principles and theory. Edw Arnold Publ Lond Marcel Dekker Inc N Y. 1965;

[10] Gonzalez FR, Alessandrini JL, Nardillo AM. Revision of a theoretical expression for gas-liquid chromatographic retention. J Chromatogr A. 1999;852(2):583–588.

[11] Ben Naim A. Solvation Thermodyanmics. Plenum Press. 1987; New York.

[12] Gonzalez FR. Interpreting the gas chromatographic retention of n-alkanes. J Chromatogr A. 2000;873(2):209–219.

[13] Aldaeus F, Thewalim Y, Colmsjo A. Prediction of retention times and peak widths in temperature-programmed gas chromatography using the finite element method. J Chromatogr A. 2009 Jan;1216(1):134–139.

[14] Gonzalez FR. Considerations on the temperature dependence of the gas-liquid chromatographic retention. J Chromatogr A. 2002;942(1–2):211–221.

[15] Castells RC, Arancibia EL, Nardillo AM. Regression against temperature of gas-chromatographic retention data. J Chromatogr. 1990;504 (1):45–53.

[16] Aldaeus F. New tools for trapping and separation in gas chromatgorahpy and dielectrophoresis (Improved performance by aid of computer simulation). Dr Thesis Anal Chem Stockh Univ. 2007;

[17] Snijders H, Janssen HG, Cramers C. Optimization of temperature-programmed

gas chromatographic separations .1. Prediction of retention times and peak widths from retention indices. J Chromatogr A. 1995;718(2):339–355.

[18] Aldaeus F, Thewalim Y, Colmsjo A. Prediction of retention times of polycyclic aromatic hydrocarbons and n-alkanes in temperature-programmed gas chromatography. Anal Bioanal Chem. 2007 Oct;389(3):941–950.

[19] Hernandez-Baez DM, Reid A, Chapoy A, Tohidi B, Bounaceur R. Establishing the Maximum Carbon Number for Reliable Quantitative Gas Chromatographic Analysis of Heavy Ends Hydrocarbons. Part 2. Migration and Separation Gas Chromatography Modeling. Energy Fuels. 2013;27(4): 2336.

[20] Davankov VA. The true physical meaning of the corrected retention volumes in GC. Chromatographia. 1997; 44(5–6):279–282.

[21] Kestin J, Knierim K, Mason EA, Najafi ST, Ro ST, Waldman M. Equilibrium and trasport properties of the noble gases and their mixtures at low density. J Phys Chem Ref Data. 1984;13(1):229.

[22] Hawkes SJ. Viscosities of carrier gases at gas-chromatograph temperatures and pressures. Chromatographia. 1993 Oct;37(7–8): 399–401.

[23] Hernandez-Baez DM, Tohidi B, Chapoy A, Bounaceur R, Reid A. Establishing the Maximum Carbon Number for Reliable Quantitative Gas Chromatographic Analysis of Heavy Ends Hydrocarbons. Part 1: Low-Conversion Thermal Cracking Modeling. Energy Fuels. 2012 May 17;26 (5):2600–2610.

[24] Golay MJE. Theory of Chromatography in Open and Coated Tubular Columns with Round and Rectangular Cross-Sections. 1958. 36 p.

[25] Blumberg LM. Temperature-Programmed Gas Chromatography. 2010.

[26] Giddings JC, Seager SL, Stucki LR, Stewart GH. Plate Height in gas chromatography. Anal Chem. 1960;32 (8):867.

[27] Snijders H, Janssen HG, Cramers C. Optimization of temperature-programmed gas chromatographic separations 0.1. Prediction of retention times and peak widths from retention indices. J Chromatogr A. 1995;718(2): 339.

Chapter 3

Features and New Examples of Gas Chromatographic Separation of Thermally Unstable Analytes

Igor G. Zenkevich

Abstract

The processes of thermal decomposition of analytes in gas chromatographic (GC) columns are classified and two new examples of them are considered in details. First of them is monomolecular decomposition of monoalkyl esters of benzene-1, 2-dicarboxylic (phthalic) acid (monoalkyl phthalates). This process has the analogy in chemical reactions in solutions and it may be responsible for the toxicity of phthalates. The second example is decomposition of non-substituted hydrazones of both aliphatic and aromatic carbonyl compounds. The analytes of the second sub-group present the first example of bimolecular (second order) decomposition in a GC column: two molecules of hydrazones form stable azines and hydrazine. Besides that this process presents the particular interest, because it is accompanied by secondary chemical reactions not in an injector, but within GC column, when a by-product of decomposition is involved into secondary interaction with other constituents of the samples. It was confirmed, that visual images of all these decomposition processes on the chromatograms are rather identical and coincide with the manifestations of interconversion of isomers or tautomers. The most often expressed features of chromatographic profiles in such cases are the presence of peaks of an initial analyte and a product of its decomposition or isomerization, connected with more or less expressed diffused "plateau" or "train" between them. The decomposition processes during sample preparation prior to chromatographic separation or in the heated injector of GC instrument are not accompanied by such features. Despite of the rather "exotic" character of the examples considered, the knowledge of them seems to be useful for better revealing the analogous situations in chromatographic practice. Thermal instability of analytes is the principal restriction of GC separation of reactive compounds and we cannot eliminate it for objective reasons. However, in some cases we can evaluate the temperature limits of chromatographic columns, which should not be exceeded during GC separation of instable compounds. The simplest (low boiling) homologs of thermally unstable compounds are often characterized by "normal" boiling point at atmospheric pressure (T_b, °C) without decomposition, that means the possibility of their GC analysis unambiguously. Therefore, we can select such T_b values as GC and/or GC–MS temperature limit (T_{lim}) for other members of series of thermally unstable homologs. If GC separation is carried out not in isothermal, but in temperature programming conditions, so-called retention temperature (T_R) of unstable analytes should not exceed the evaluated T_{lim} value.

Keywords: gas chromatography, GC-MS, thermally unstable analytes, new examples of thermal instability, monoalkyl phthalates, non-substituted hydrazones, secondary reactions in GC column, stability of analytes criterion

1. General comments (introduction)

Possible instability of analytes prior or during their separation seems to be the key restriction of chromatographic methods, because it distorts the analytical results and can make them even inacceptable at all due to their irreproducibility and/or incorrectness. The influence of instability due to the different reasons can manifest itself at different stages of analytical procedures, hence it may be considered as a way of classifying them.

The instability of components in samples prepared for analysis (**reason I**). Besides the immanent chemical instability of analytes, it can be possible due to their oxidation by air oxygen or hydrolysis by impurities of water in air or in a solvent. These reasons can be avoided by special sample protection that was demonstrated for chemically high reactive fluoro- and chlorosilanes (hydrolysis), boranes, germanes (oxidation), etc. [1]. Gas Chromatographic Retention Indices (**GC RI**) of such labile compounds determined even at the late 1970s [1] remained unique up to present. The instability during GC separation can be caused by *thermal degradation of unstable analytes in a heated injector* (**reason II**), as well. The similar **reason III** of instability seems to be the *mutual interaction of most reactive components of the samples with each other at the temperatures of GC injection*, if even their mixtures are stable in the samples prepared at the ambient conditions. However, the decomposition processes in the injector can be, if not eliminated, then controlled and sometimes minimized by varying its temperature, as well the application of so-called on-column injection. On the other side, *the decomposition processes within chromatographic columns* (**reason IV**) are usually much more difficult to manage and they represent the most difficulties in chromatographic practice. Often enough, in the result of such processes the analytical parameters (e.g., RIs) of decomposition products are attributed to the initial unstable analytes that are the typical examples of misidentification.

In high performance liquid chromatography (HPLC) the main reason of instability is the interaction of analytes with components of an eluent (most often hydrolysis).

Sometimes the revealing of analytes' instability seems to be not so simple. The main signs of analytes' instability are inconstancy of absolute or relative areas of some chromatographic peaks (not all of them) depending on variations of analytical parameters (at first, injector or column temperatures), appearance of additional peaks, distortions of peaks' shapes, loss of separation efficiency, etc. The lack of reference RI values for any reactive compounds in contemporary mass spectrometric (**MS**) and gas chromatographic (**GC**) databases (e.g., [2]) is often due to just their instability. So-called analytical artifacts (when the results do not match the analytes containing in the samples) summarized by Middleditch [3] are often caused by the mentioned reasons.

2. Different kinds of analytes' instability complicating gas chromatographic analysis (like their classification)

As the typical examples of the manifestation of instability in the result of high reactivity of analytes (**reason I**) such semi-volatile compounds as **(3-aminopropyl) trimethoxy-** (boiling point, T_b, 194°C) and **(3-aminopropyl)triethoxysilane**

(T_b 220 ± 3°C) [4, 5] can be mentioned. It is important to note that these compounds have the normal boiling points at atmospheric pressure, but they are as active as silanization agents that can react even with glass surfaces of chromatographic syringes used for injection of samples and the silica surfaces of injector liners and chromatographic columns. Due to this reason these compounds remain uncharacterized by GC RIs up to date [2]. Another example of the same kind is so "exotic" compound as **dimethyl thionitrosamine, $(CH_3)_2N\text{-}N{=}S$** that is so unstable as the individual substance or constituent of concentrated solutions at the ambient temperature due to its easy polymerization, that it exists only in dilute solutions [6]. At the same time, this compound seems relatively stable within a chromatographic column, so that its RI on standard non-polar stationary phase has been successfully determined (992 ± 2)[1] [6].

The examples of analytes' thermal decomposition in a heated injector (**reason II**) are numerous, as well. Thus, over a dozen publications presenting RI values of **N-nitrosodiphenylamine, $(C_6H_5)_2N\text{-}NO$**, are known up to present, but only one of them contains the correct RI value 1865 [7], while others belong to the decomposition product – diphenylamine, **$(C_6H_5)_2NH$** (RI 1587 ± 13 [2]). For such decomposition to take place a source of active hydrogen atom is required; it can be using of hydroxyl-containing solvents, or water residues in a sample.

Sometimes the revealing of structural features of analytes' molecules responsible for their decomposition appears the important problem in organic chemistry. Thus, such heterocyclic compounds as 4-acyl-1,3,4-oxadiazolines do not differ principally from other organic compounds by their stability. Nevertheless, if these heterocycles contain no substituents in the position 2, they decompose at the injector temperatures above 150°C with formation of monoacyl hydrazones [8] in the result of decarbonylation. Within temperature range 150–190°C both initial compounds and decomposition products are detected, but above 190°C no peaks of initial analytes are registered on the chromatograms:

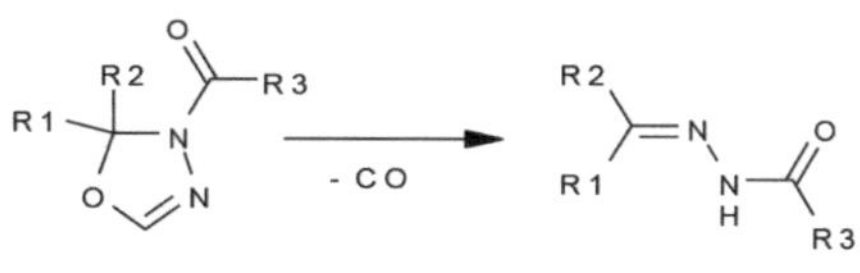

The similar decomposition is observed for substituted high reactive 3,4-dihydroformazans (trivial name "azohydrazines", but the limit of their thermal stability is less, approx. not more than 130°C [9]. The decomposition products are disubstituted hydrazones:

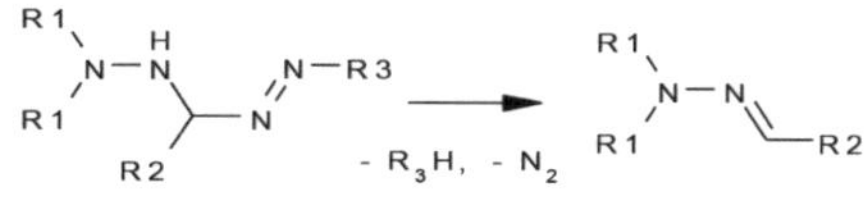

As an example of erroneous determination of retention indices in the result of thermal decomposition of analytes let us mention the RI value for trichloroacetic acid (600) on standard non-polar polydimethylsiloxane stationary phases published in Sadtler Retention Index Library [10]. It looks like obviously erroneous, but its proving appeared to be not so simple, because it requires comparing RI values for series of structural analogues of acetic acid containing up to two chlorine atoms (congeners), as it is illustrated by data in **Table 1**:

[1] Here and afterwards all RI values, if it is not mentioned specially, presented for standard non-polar stationary phases (polydimethyl siloxanes), i.e., $RI_{non\text{-}polar}$.

Acid congener	Structure	Boiling point, °C	RI ± s_{RI}
Acetic	CH_3CO_2H	118	624 ± 10 [2]
Chloroacetic	$ClCH_2CO_2H$	188	864 ± 12*
Dichloroacetic	Cl_2CHCO_2H	194	1048 ± 23*
Trichloroacetic	Cl_3CCO_2H	197; Extrapolated value is 194.6°C**	No experimental data; extrapolated value is 1189**

**RI values for chloro- and dichloroacetic acids are taken from author's data collection.*
***T_b and RI values are extrapolated using the data for previous congeners and the recurrent algorithm [11].*

Table 1.
Comparing the retention indices ($RI_{non\text{-}polar}$) for chloroacetic acids.

Known RI data for substituted acids Cl_0-Cl_2 (three points) allows approximating RIs using recurrent relation $\mathbf{RI}(n_{Cl}+1) = a\mathbf{RI}(n_{Cl}) + b$ [11] (a = 0.767, b = 385.6), hence for trichloroacetic acid (Cl_3) RI = 0.767 × 1048 + 385.6 ≈ 1189. Thus, the value 600 from [10] is erroneous and obviously belongs to the chloroform $CHCl_3$ (reference RI value is 605 ± 4 [2]) formed in the result of decomposition of the trichloroacetic acid within heated GC injector:

$$Cl_3C-CO_2H \rightarrow CHCl_3 + CO_2$$

Thermal instability caused by mutual interaction of constituents of the samples in an heated injector (**reason III**) can be illustrated by features of chromatographic determination of the impurity of 1,2-propanediol (propylene glycol) in the high boiling polar aprotic organic solvent – 4-methyl-1,3-dioxolan-2-one (trivial name – propylene carbonate, T_b 242°C) [12]. The real content of propylene glycol in this solvent at the ambient temperature is strongly distorted in the result of its hydrolysis by the residual amounts of water in a heated injector during injection:

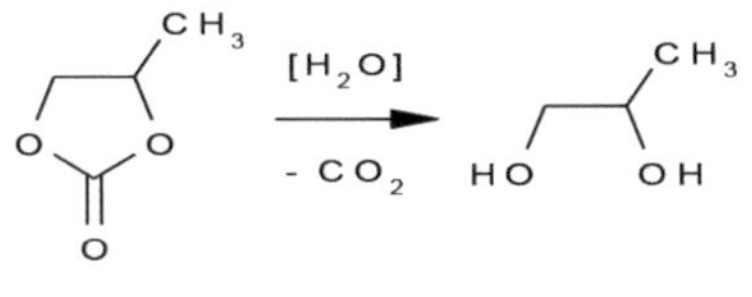

The last source of instability of analytes during GC separation – decomposition of analytes in a chromatographic column (**reason IV**) – is not equivalent to their decomposition in an injector (**reason II**). In a column analytes are usually exist under the influence of a lower temperature, but a longer time than in an injector. We cannot change the influence of column's temperature without changing the retention time (t_R), so far as these parameters are connected by the known two-parameters Antoine-like equation:

$$\log(\boldsymbol{t}_R - \boldsymbol{t}_0) = \boldsymbol{a}/\boldsymbol{T} + \boldsymbol{b} \quad (1)$$

where t_0 is hold-up time of chromatographic system, T is the absolute temperature (in Kelvins), coefficients a and b are calculated by Least Squares Methods.

The temperature dependence of the rate of first-order decomposition reaction (rate constant k) is approximated with known Arrhenius relation that is two-parameter Antoine-like equation as well:

$$\ln \boldsymbol{k} = -\boldsymbol{E}_a / \boldsymbol{RT} + \ln \boldsymbol{A} \quad (2)$$

where E_a is activation energy for the reaction, R is universal gas constant, A is pre-exponential factor.

Thus, the influence of column's temperature on the areas of chromatographic peaks of unstable compounds is simultaneously determined by four coefficients, a, b (Eq. 1), E_a/R, and ln$A$ (Eq. 2). Just these relationships confirm that analytes' decomposition in GC column seem to be difficult to eliminate, because if we decrease the column's temperature, we increase the analyte's residence time in the heated area. Besides the examples listed above, the decomposition inside GC column (in addition to decomposition in an injector) is valuable for diazocarbonyl compounds, decomposed with the formation of substituted ketene in the result of so-called Wolff rearrangement [13]:

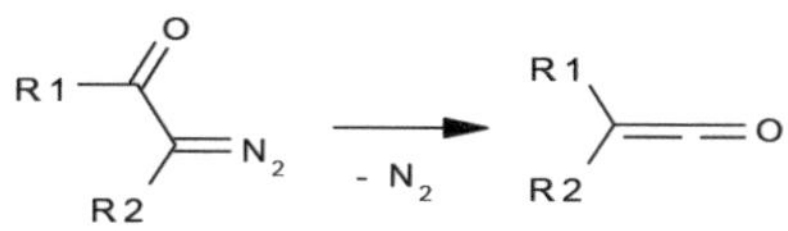

It is interesting to note that resulted ketenes are unstable compounds as well; they cannot be isolated as individual substances or components of concentrated solutions due to easy polymerization, and, hence, do not form distinct chromatographic peaks. In the result, their mass spectra were registered successfully [13], but GC RIs cannot be determined excepting the most volatile simplest member of this series – dimethyl ketene ($RI_{non\text{-}polar}$ 484, RI_{polar} 1215 [2]).

3. The influence of the decomposition of analytes in chromatographic columns on contours of chromatograms

The last mentioned mode of analytes' transformations during GC separation (**reason IV**, decomposition in a chromatographic column) is often manifested in the appearance of specific profiles of chromatograms. **Figure 1** presents a schematic image of typical chromatogram of unstable analyte "*X*" which is converted into decomposition product "*Y*" within a chromatographic column in the result of the process $X \rightarrow Y$. Usually (at least, in GC) the decomposition product "Y" is more

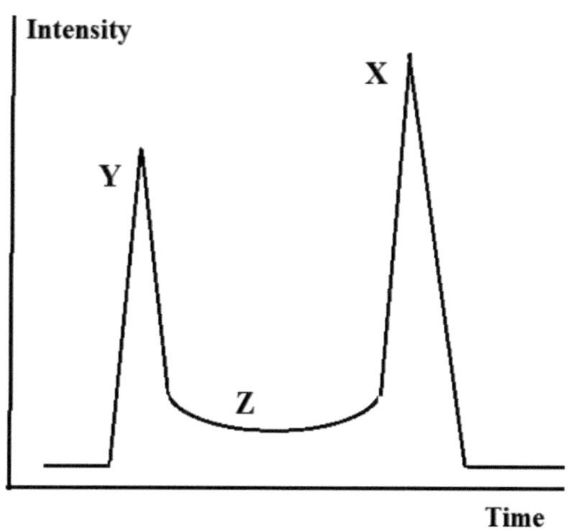

Figure 1.
Schematic image of the typical chromatogram of thermally unstable analyte "X" converted into more volatile decomposition product "Y" within chromatographic column during separation. The appearance of a diffuse zone "Z" ("train" or "plateau") between peaks "X" and "Y" is the sign of decomposition (X → Y).

volatile simpler compound. The peaks of both compounds "*X*" and "*Y*" rather often can be registered on the chromatograms; and retention times of decomposition products "*Y*" are naturally less than those of initial analytes "*X*". However, besides these two regular peaks there is a specific diffuse signal between them (*Z*), named as "anomalous profile", "plateau", "train", "plume", or something similar (no generally accepted terminology exists). In the case of monomolecular decomposition of *X*, this "train" consists exclusively of decomposition product "*Y*".

The "idealistic" profile like that on **Figure 1** in many cases can be more or less distorted that complicates revealing the decomposition processes in a column. Some examples of such distortions are presented on Figures below. For example, decomposition of **1-diazo-4-phenylbutan-2-one, $C_6H_5CH_2CH_2COCH{=}N_2$** within a column leads to the formation of "train" (*Z*) of ketene **$C_6H_5CH_2CH_2CH{=}C{=}O$** prior the peak of initial diazocompound, but this ketene does not form a separate chromatographic peak [13].

The similar profiles of chromatograms are observed not only in cases of decomposition of analytes (irreversible reactions), but in cases of reversible interconversion of their isomeric or tautomeric forms. It should be noted that similar profiles may be observed both in GC, and HPLC. Such chromatograms were registered for keto- and enolic tautomers of 1,3-diketones (GC) [14], β-ketoesters (GC) [15, 16], *syn*- and *anti*-isomers of 2,4-dinitrophenyl hydrazones of carbonyl compounds (HPLC) [17, 18], derivatives of hydroxyquinones (HPLC) [19], and so on. The fragment of a GC–MS TIC-chromatogram illustrating the separation of keto- and enolic tautomers of ethyl acetoacetate $CH_3COCH_2CO_2C_2H_5$ is presented at **Figure 2**. The "train" observed here between the peaks of tautomers is similar to a schematic profile at **Figure 1**.

Comparing the examples mentioned we can conclude that the decomposition of unstable or reactive analytes prior to or during the chromatographic separation restricts the possibilities of this analytical technique and complicates the interpretation of results. Due to this reason any new examples of such decomposition should be reliably revealed and discussed to avoid difficulties in the subsequent data interpretation. According with this viewing, two new examples are considered here in details, namely unusual thermal instability of monoalkyl esters of benzene-1,2-dicarboxylic acid (monoalkyl phthalates) and decomposition of non-substituted hydrazones of carbonyl compounds.

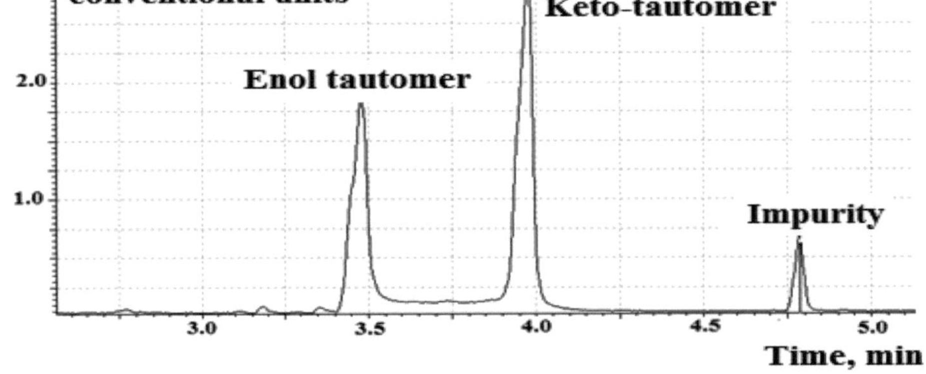

Figure 2.
*The fragment of the Total ion current (**TIC**) chromatogram (WCOT column with standard non-polar polydimethyl siloxane stationary phase) containing the peaks of keto and enol tautomers of ethyl acetoacetate* **$CH_3COCH_2C_2H_5$**. *The existence of "plateau" between peaks of tautomers confirms their interconversion within GC column during separation. Reproduced from [16] with permission.*

4. Materials and methods

Most of chemicals (alcohols, carbonyl compounds, and ethyl acetoacetate) of reagent or chromatographic grade were from "Reakhim" (Moscow, Russia). 2-Propanol and methylene chloride of chemical grade (solvents) were provided from "Vekton" (St. Petersburg, Russia). Hydrazine hydrate of reagent grade was purchased from Acros Organics, Belgium.

4.1 Preparation of reaction mixtures

Monoalkyl phthalates. Approx. 6 mg of phthalic anhydride (melting point 130–132°C, approx. 50 μmol) was added to the 2-mL portions of 1-alkanols $C_nH_{2n+1}OH$ (n = 1–8) and heated in the presence of catalytic amounts of phosphorous acid at the boiling point of the alcohol (for alcohols $C_1 - C_4$) or not more than 110°C (for alcohol $C_5 - C_8$) during 30 min to complete the dissolution of the phthalic anhydride. For GC–MS analysis 5 μL of reaction mixtures were diluted with 0.5–2.5 mL of methylene chloride by a factor of 100–500. All monoalkyl phthalates are characterized without their isolation from reaction mixtures. The list of RI values for monoalkyl phthalates is presented in **Table 2**.

Non-substituted hydrazones of carbonyl compounds. Hydrazine hydrate (2 mL) were mixed with 50 μL of carbonyl compounds (molar excess from 60:1 to 120:1) and 2 mL of 2-propanol at the ambient temperature. To increase the yields of azines, if necessary in some experiments, the molar ratio of hydrazine was decreased to 30:1 and 15:1. After 10 min 50 μL of obtained mixtures were diluted with 2 mL of 2-propanol, followed by addition of 2 μL of the reference *n*-alkanes mixture. All hydrazones were characterized without isolation from reaction mixtures. Besides hydrazones these mixtures contained variable amounts of azines. The list of RI values for non-substituted hydrazones and corresponding azines is presented in **Table 3**.

R in $C_6H_4(CO_2H)(CO_2R)$	*M*	RI [2]	RI [20]	R in $C_6H_4(CO_2H)(CO_2R)$	*M*	RI [2]	RI [20]
CH_3	180	1530*	1566 ± 18	*neo*-C_5H_{11}	236	1809	—
C_2H_5	194	1651	1641 ± 5	C_6H_{13}	250	2023	2029 ± 6
C_3H_7	208	1731	1734 ± 6	$-CH_2CH(C_2H_5)_2$	250	1977	—
iso-C_3H_7	208	1667	1666 ± 2	C_7H_{15}	250	2128	2132 ± 6
C_4H_9	222	1828	1849 ± 11	2-C_7H_{15}	264	—	2017 ± 10
iso-C_4H_9	222	1771	1777 ± 3	C_8H_{17}	278	2236	2250 ± 17
2-C_4H_9	222	1764*	1758 ± 3	2-C_8H_{17}	278	2143	2134
C_5H_{11}	236	1926	1929 ± 11	$-CH_2CH(C_2H_5)C_4H_9$	278	2152	—
iso-C_5H_{11}	236	—	1890 ± 7	C_9H_{19}	292	2325	—
2-C_5H_{11}	236	1842	1844 ± 2	$C_{10}H_{21}$	306	2431	—

Not experimental, but RI values evaluated using additive scheme [2] are marked with asterisk

Table 2.
Gas chromatographic retention indices of some monoalkyl phthalates on semi-standard non-polar polydimethylsiloxane stationary phases (95% methyl and 5% phenyl groups).

Carbonyl compound	RI of hydrazone	RI of azine
Acetone	750 ± 6	848 ± 3
3,3-Dimethyl- 2-butanone	936 ± 2	—
2-Hexanone	1054 ± 3	1357 ± 4 1394 ± 4*
4-Methyl-2-pentanone	970 ± 3	1249 ± 3 1298 ± 3*
2-Heptanone	1163 ± 3	1552 ± 4
3-Heptanone	1118 ± 3	1471 ± 3 1506 ± 5*
2-Octanone	1272 ± 2	1721 ± 3
Cyclopentanone	1032 ± 2	1470 ± 3
Cyclohexanone	777 ± 2	1638 ± 5
Acetophenone	1382 ± 4	2082 ± 5
4-Phenyl-2-butanone	1454 ± 3	2168 ± 5

**The minor constituents with mass spectra identical to spectra of principal (E,E)-isomers tentatively belong to (E,Z)-isomers of azines.*

Table 3.
Gas chromatographic retention indices of non-substituted hydrazones and corresponding azines on semi-standard non-polar polydimethylsiloxane stationary phases [24].

4.2 Instrumentation and data processing

GC–MS analyses of reaction mixtures were performed using Shimadzu QP 2010 SE gas chromatograph – mass spectrometer with electron ionization (70 eV) equipped with RTX-5 MS column of the length 30 m, internal diameter 0.32 mm, and stationary phase film thickness 0.25 μm. The conditions of analysis were as follows: temperature programming regime, initial temperature 70°C (phthalates) or 50°C (hydrazones), and ramp 6 K/min (phthalates) or 10 K/min (hydrazones) up to 200°C. Helium was used as carrier gas, flow rate was 1.8 mL/min, split ratio 1:10–1:12, injector temperatures were 200–250°C, interface and ion source temperatures were 200°C. Volumes of injected samples were 0.5 μL (phthalates) or 2 μL (hydrazones). To determine retention indices, mixtures of $C_7 - C_{20}$ *n*-alkanes (in different combinations) were added to the samples before analysis for determination of retention indices (RI) (calculations of linear or linear-logarithmic RIs were provided using home-made QBasic program).

Statistical data processing and plotting of results were carried out using the Origin software (versions 4.1 and 8.2). The task considered requires the specific mode of results' presentation, preferably in the form of chromatogram images. Most of original chromatograms besides the peaks of target analytes and decomposition products contain peaks of reference *n*-alkanes, which were required for calculation of retention indices and confirmation of the appropriate effectiveness of a column.

5. Results and discussion

5.1 Features of GC separation of monoalkyl phthalates

Contemporary GC–MS data bases (e.g., like [2]) provide important information for identification of analytes including standard electron ionization (EI) mass spectra and GC retention indices (**RI**) on standard non-polar and polar phases. However, besides that, considering the large collection of reference data allows revealing both individual compound and their various taxonomic sub-groups (homologous series, multitudes of homologs and/or congeners, etc.) insufficiently characterized by analytical parameters up to present. One of such groups appeared to be the acidic esters of polycarboxylic acids, including alkyl esters of benzene-1,2-dicarboxylic acid (monoalkyl phthalates) that stimulated determination of their MS and GC analytical parameters [20] in comparison with the data for their much better characterized structural analogues – dialkyl phthalates. Rather unexpectedly it was found that monoalkyl phthalates appeared to be unstable at standard conditions of GC separation.

Monoalkyl phthalates can be easily synthesized from phthalic anhydride and corresponding alcohols at acid catalysis in accordance with the following scheme:

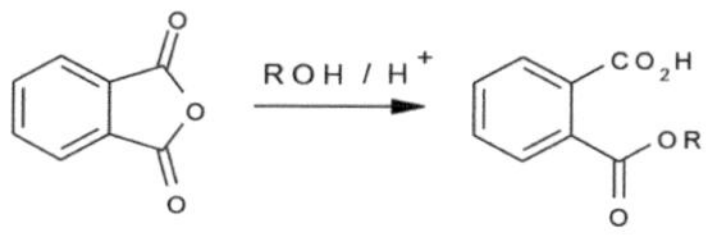

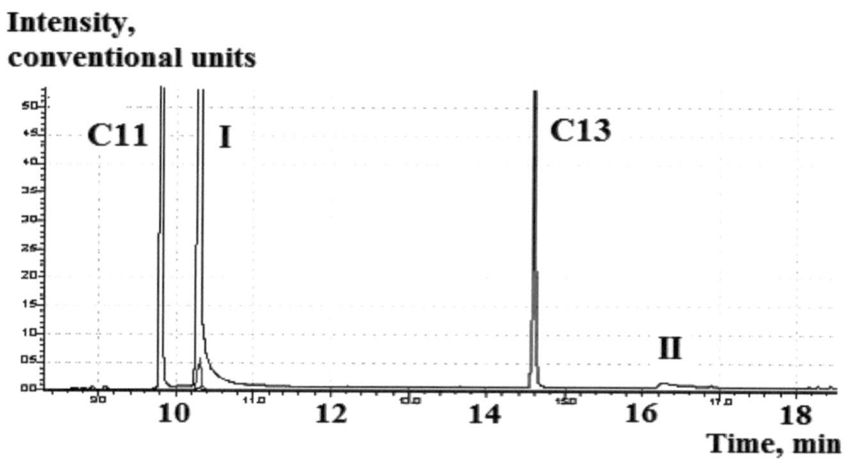

Figure 3.
The fragment of the TIC-chromatogram of reaction mixture of phthalic anhydride with methanol at the comparable scale of peak intensities. Peak (**I**) *– Phthalic anhydride, peak* (**II**) *– Monomethyl phthalate,* C_{11} *and* C_{13} *– Peaks of reference n-alkanes. Reproduced from [20] with permission.*

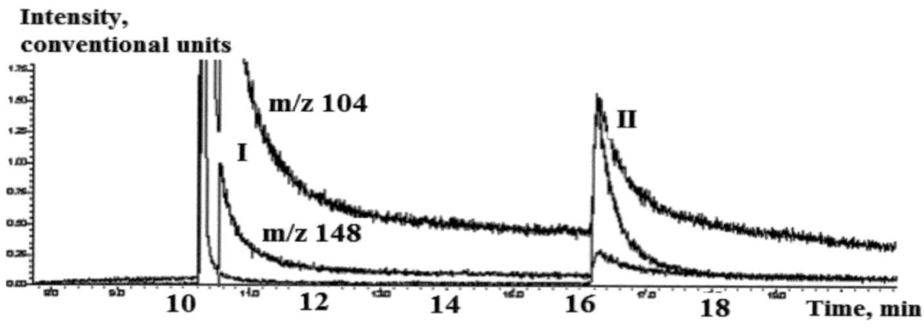

Figure 4.
The same fragment of the chromatogram of reaction mixture of phthalic anhydride with methanol as those at **Figure 3** *in SIM-regime on a larger scale; peak* (**I**) *– Phthalic anhydride, peak* (**II**) *– Monomethyl phthalate, m/z 148 – Molecular mass of phthalic anhydride, m/z 104 – Maximal signal in the mass spectrum of phthalic anhydride. Reproduced from [20] with permission.*

The large molar excess of the alcohols in the reaction mixtures allowed us to hope for a high degree of conversion of phthalic anhydride into monoalkyl phthalates. Surprisingly all the chromatograms of reaction mixtures indicate the predominant amounts of phthalic anhydride. For instance, the fragment of the TIC-chromatogram of the reaction mixture of phthalic anhydride with methanol is presented on **Figure 3**; the ratio of peak areas of anhydride (**I**) to monomethyl phthalate (**II**, RI is 1566 ± 18, retention temperature of 152°C) is approx. 100:1. To explain such an anomaly, the original chromatogram was reconstructed into SIM-mode using values *m/z* 148 (molecular mass of phthalic anhydride) and *m/z* 104 (maximal peak in its mass spectrum); the result is shown at **Figure 4**.

From **Figure 4** it is easy to notice that both signal with *m/z* 104 and *m/z* 148 besides the peak of phthalic anhydride itself indicate the second local maxima for monomethyl phthalate (**II**). However, the most important information from this SIM-chromatogram is revealing the registered "train" between peaks (**I**) and (**II**). It is the unambiguous indication of thermal instability of monomethyl phthalate at conditions of GC separation, and its main decomposition product is phthalic anhydride. The same process can be assumed for other monoalkyl phthalates having the higher boiling points than monomethyl ester, and, hence, the higher retention temperatures [20]:

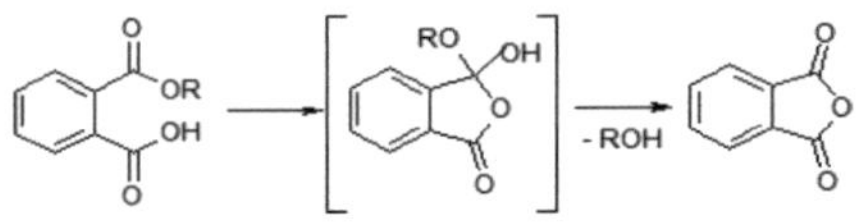

A similar process of cyclization with participation of two carbonyl groups in the *ortho*-position in benzene ring is known for chemical reaction in solution. It explains, for instance, the formation of 6-chloro-3-methoxyphthalide from 2-acetyl-5-chlorobenzoic acid [21]:

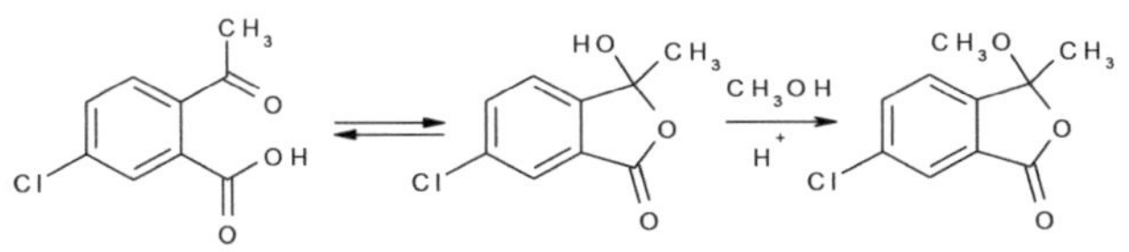

The additional independent confirmation of the thermal decomposition of monomethyl phthalate during gas chromatographic separation its mass spectrum from the database [2] should be considered. At **Figure 5** the mass spectrum of monomethyl phthalate (a) is compared with that of phthalic anhydride (b). The similarity of positions and intensities of main peaks of these spectra with *m/z* 104, 76, and 50 looks noteworthy. Most intensive peak of monomethyl phthalate itself belongs to the ions $[M - CH_3O]^+$ with *m/z* 149, but its relative intensity in mass spectrum [2] is about 40%. Moreover, the library search (reversed mode) for monomethyl phthalate [2] gives only two results with matching factor Q > 0.800, namely the same ester (another mass spectrum) with Q = 911, and phthalic anhydride (Q = 807). The similar paradox is observed even for phthalic anhydride itself; one of the results of the library search for its mass spectrum (in reversed mode) is monoethyl phthalate with Q = 854.

Nevertheless, more preferable mass spectrum of monomethyl phthalate can be obtained in the result of the accurate subtracting of background and/or overlapping signals. So as not to overload the text with figures, let us list it in the numerical form, $m/z \geq 39$ ($I_{rel} \geq 2\%$), **M** is the symbol of molecular ions:

180(2)**M**, 163(2), 150(9), 149(100) $[M - CH_3O]$, 148(10), 137(4), 136(31), 135(18), 122(5), 121(20), 118(2), 106(5), 105(52) $[M - CH_3O - CO_2]$, 104(65)

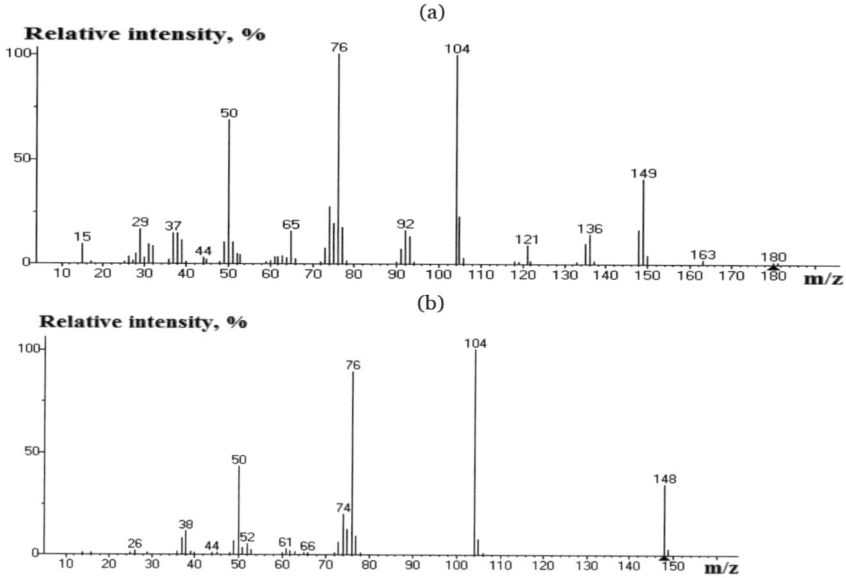

Figure 5.
Comparison of mass spectra of monomethyl phthalate (a) and phthalic anhydride (b) from database [2]. The mass spectrum of monomethyl phthalate is distorted by mass spectrum of phthalic anhydride.

[C_7H_4O], 94(2), 93(29), 92(34), 91(16), 78(4), 77(33), 76(76) [C_7H_4O – $CO \equiv C_6H_4$], 75(14), 74(21), 73(5), 71(3), 66(5), 65(37), 64(5), 63(5), 59(3), 58(2), 57(2), 53(3), 52(6), 51(16), 50(46) [C_4H_2], 49(4), 44(3), 43(3), 41(2), 39(24).

As it can be seen, this mass spectrum strongly differs with that from database [2]; the maximal peak appears to be *m*/*z* 149 as it should be for esters of phthalic acid [22]. However, the presence of the signals with *m*/*z* 104, 76, and 50 do not allow the excluding it's at least partial decomposition in a GC column or at any point between column and ion source of mass spectrometer.

The instability of monoalkyl phthalates – the products of partial hydrolysis of dialkyl phthalates – widely used plasticizers of polymeric compositions – permits us to suggest the novel interpretation of endocrinic toxicity of these esters. Their decomposition in small extent can take place not only at the heating, but at the ambient conditions, as well. The product of such decomposition – phthalic anhydride – is an active acylation reagent which can react with some targets inside the living cells (e.g., peptides or nucleic acids) [20].

5.2 Anomalous chromatographic properties of non-substituted hydrazones of carbonyl compounds

The considering of database [2] allows revealing another series of simple organic compounds that are characterized in enough extent neither mass spectra, nor GC retention indices. It is the products of the nucleophylic addition of hydrazine to carbonyl compounds – non-substituted hydrazones. These compounds are used in practice of organic synthesis since 19th century:

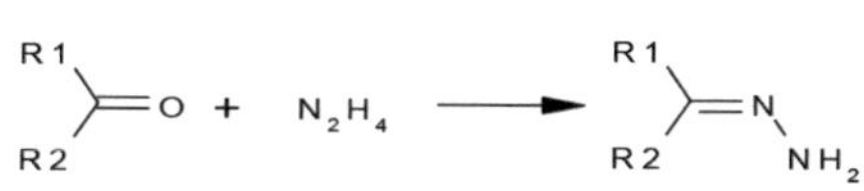

Such hydrazones can be synthesized from carbonyl compounds in one stage; they are intermediates in so-called Wolff-Kishner reduction of carbonyl

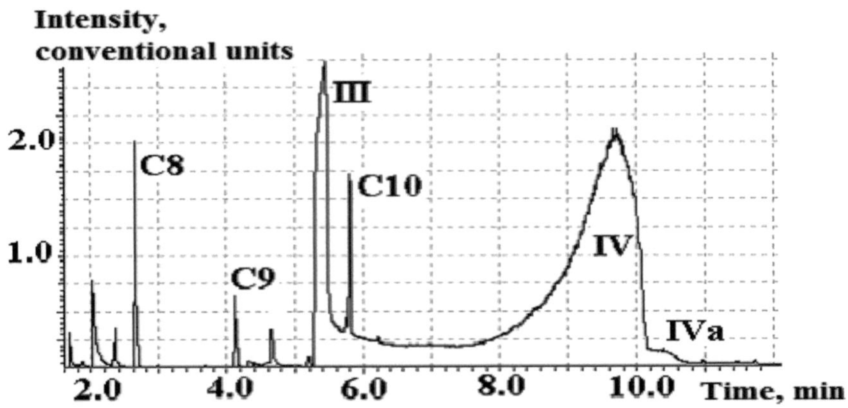

Figure 6.
The fragment of the TIC-chromatogram of reaction mixture of 4-methyl-2-pentanone with hydrazine hydrate; peak (**III**) *– 4-methyl-2-pentanone hydrazone, peak* (**IV**) *– 4-methyl-2-pentanone azine,* C_8 *–* C_{10} *– Reference n-alkanes. Reproduced from [24] with permission.*

compounds into hydrocarbons with the same carbon skeletons [23]. Under such circumstances it seems nearly paradoxical why these synthetically important compounds remained not characterized both by mass spectra and by gas chromatographic analytical parameters until last time. The database [2] contains mass spectrum and RI value for only one simplest member of this series – acetone hydrazone **$(CH_3)_2C$=N-NH_2**.

Such inconsistency explains the necessity to carry out GC–MS analysis of reaction mixtures of carbonyl compounds with hydrazine hydrate [24]. Surprisingly, instead of "normal" (more or less sharp) chromatographic peaks of hydrazones the very "blurry" signals were recorded for them, as it can be seen from fragment of TIC-chromatograms of the reaction mixtures of hydrazine hydrate with 4-methyl-2-pentanone (**Figure 6**). The chromatograms of reaction mixtures of other carbonyl compounds look similar (see ref. [24]). Any attempts to improve the shapes of chromatograms by varying separation conditions or by dilution of samples remained unsuccessful.

All the chromatograms contain two diffuse peaks with variable relative intensities connected by "trains" between them. Strong broadening the peaks explains us the low accuracy of determining the retention indices of reaction products (and their low interlaboratory reproducibility, as well). From mass spectrometric data it is easy to conclude that molecular masses of the first eluted peaks correspond to non-substituted hydrazones, while those of the latter peaks – to azines of carbonyl compounds – stable products of the bimolecular decomposition of non-substituted hydrazones:

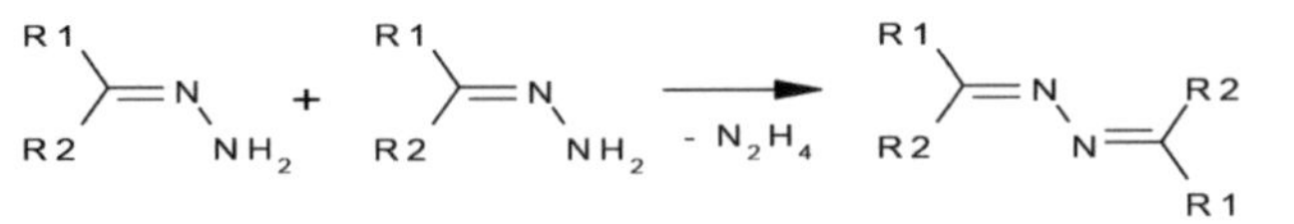

Besides this way of formation, more stable azines are the "normal" by-side reaction products even at the large excess of hydrazine. Mass spectra registered in various points between hydrazones and azines indicate that the "trains" between them are formed both by hydrazones and azines in variable proportions. It is the principal difference of the nature of such "trains" for mono-molecular decomposition processes in a GC column, when these areas are formed by decomposition products only.

The decomposition of just non-substituted hydrazones in a GC column is confirmed by analysis of reaction mixtures containing solely azines. For instance, the chromatograms of the reaction mixtures of cyclohexanone with hydrazine hydrate in 10 minutes after mixing the reagents and after one week storage this sample at the ambient temperature look rather different. On the first chromatogram the intensive peak of hydrazone is observed, whereas prior to the peak of azine there is the "train" confirming decomposition process in a GC column. One week later when the sole reaction product appeared to be the azine; the "train" before its peak is disappeared completely.

The decomposition of monoalkyl phthalates considered in the Section 3.1 like other known decomposition processes in a chromatographic column is the first order reactions (monomolecular). On the contrary, the decomposition of non-substituted hydrazones with formation of azines is the first revealed example of second order (bimolecular) reaction in gas chromatographic column. It should be specially noted that the observed features of chromatograms ("trains" or "plateau") in the case of bimolecular decomposition of non-substituted hydrazones are the same, as those in the cases on monomolecular reactions.

5.3 Decomposition followed by secondary chemical reactions in GC column

After examples presented above we can consider the most unusual example of analytes' conversion in a GC column. **Figure 7** contains the fragment of the TIC-chromatogram of aromatic carbonyl compound – acetophenone – with hydrazine hydrate. Similarly to the previous examples we observe the peaks of acetophenone hydrazone (**IX**, t_R approx. 11.5 min), acetophenone azine (**X**, t_R approx. 19.0 min) connected by a "train" between them (zone **Z1**), as well the peak of initial acetophenone (t_R approx. 7.5 min). However, besides the expected zone **Z1**, the second anomalous area **Z2** is observed prior to the peak of hydrazone (**IX**). Mass-spectra registered in the different points of this area **Z2** indicated that it composed exclusively of acetophenone hydrazone with the following mass spectrum in the numerical form, $m/z \geq 39$ ($I_{rel} \geq 2\%$):

135(10), 134(100)**M**, 133(21), 120(4), 119(42) [M – CH_3], 118(5), 117(20) [M – NH_3], 104(3), 103(12), 102(3), 93(9), 92(10), 91(4), 90(2), 89(2), 79(2), 78(10), 77(88) [C_6H_5], 76(7), 75(3), 74(3), 66(4), 65(6), 63(4), 57(9), 56(5), 52(3), 51(19).

The paradox observed is the follows: the area **Z2** corresponds to the range of retention times which are less than retention time of "normal" acetophenone hydrazone. In other words, *some part of acetophenone hydrazone is eluted from GC column*

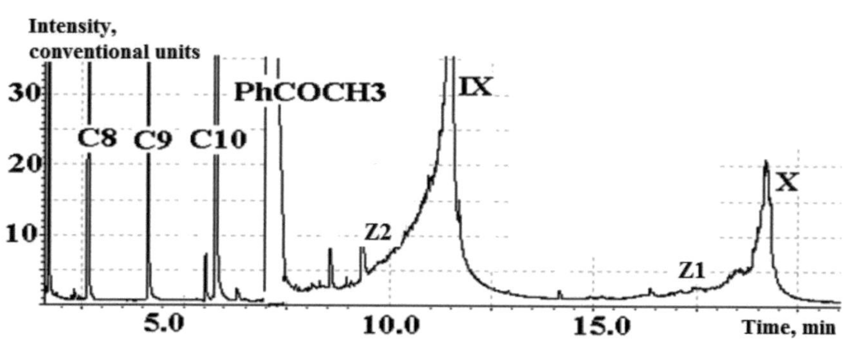

Figure 7.
The fragment of the TIC-chromatogram of the reaction mixture of acetophenone with hydrazine hydrate; peak (IX) – Acetophenone hydrazone, peak (X) – Acetophenone azine, C_8 – C_{10} – Peaks of reference n-alkanes. Reproduced from [24] with permission.

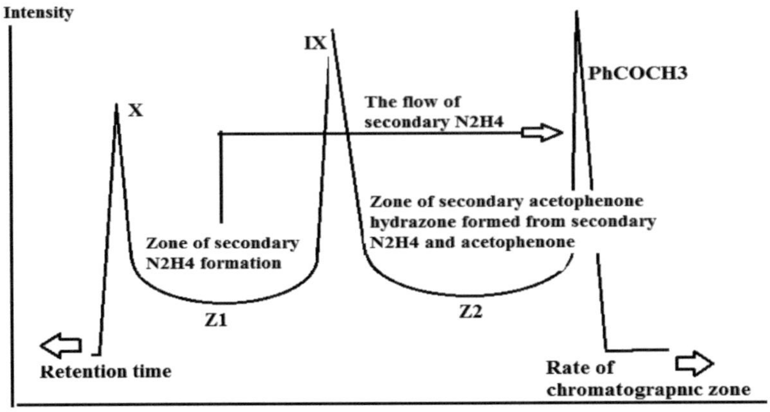

Figure 8.
The scheme of formation of two "trains" on chromatograms. Zone **Z1** *– Area of acetophenone hydrazone* (**IX**) *decomposition with the formation of acetophenone azine* (**X**) *and secondary hydrazine; zone* **Z2** *– Area of interaction of secondary hydrazine with residues of initial acetophenone in reaction mixture with formation of secondary acetophenone hydrazone been eluted before "normal" acetophenone hydrazone.*

"before" acetophenone hydrazone (!). Even the very formulation of this effect looks nearly paradoxical. It is the first known example of so anomalous chromatographic behavior of analytes.

To explain this anomaly let us reconstruct schematically the TIC-chromatogram of reaction mixture of acetophenone with hydrazine hydrate in the reversed scale, like it is shown at **Figure 8**. At this scheme an increase in retention times corresponds to a direction to the left, while an increase in the rate of chromatographic zones within column – to a direction to the right. The decomposition of acetophenone hydrazone (**IX**) with formation of acetophenone azine (**X**) (according with the scheme of reaction above) is occurs in zone **Z1** and it is accompanied by formation of secondary hydrazine that is most volatile constituent comparing with other reaction products (T_b 114°C). Thus, the rate of chromatographic zone of secondary hydrazine exceeds the rates of chromatographic zones of other components of reaction mixtures. This hydrazine zone can "catch up" the zone of initial acetophenone and reacts with it, that leads to the formation of the "train" of secondary acetophenone hydrazone (**Z2**) located *between* the peaks of acetophenone itself and the "normal" peak of acetophenone hydrazone (i.e., before it). Of course, this reaction of nucleophylic addition of hydrazine to the carbonyl compound proceeds not in the gaseous, but in the condensed phase, namely in the stationary phase film.

Despite of the highly "exotic" character of the examples considered, the knowledge of them seems to be useful for better revealing the analogous situations in routine chromatographic practice.

5.4 Criterion of applicability of GC / GC-MS analysis for thermally unstable compounds

Obviously, thermal instability of analytes seems to be the principal restriction of gas chromatographic separation of highly reactive compounds, thus we cannot eliminate it for objective reasons. However, in some cases we can evaluate the temperature limits of chromatographic columns, which should not be exceeded when analyzing certain unstable analytes.

It is known that the low boiling simplest homologs of thermally unstable compounds can often be distilled at ambient conditions without their thermal decomposition and, therefore, they are characterized by "normal" boiling point at atmospheric pressure (T_b, °C). The existence of boiling point means the possibility of gas chromatographic analysis of such compounds using contemporary fused silica WCOT columns of high inertness. Heating the higher members of the series to their boiling points at atmospheric pressure appeared to be impossible because of their decomposition. In practice of organic synthesis, distillation of such compounds under reduced pressure is used. Thus, we can select the boiling point of simplest homolog of the series under consideration at atmospheric pressure as the GC and/or GC–MS analyses temperature limit (T_{lim}) for other members of these homologous series. If GC separation is carried out not in isothermal, but in temperature programming conditions, instead of fixed column's temperature we must operate with so-called retention temperature (T_R) which should not exceed T_{lim} value:

$$\boldsymbol{T}_{\mathrm{R}} = \boldsymbol{T}_0 + \boldsymbol{r t}_{\mathrm{R}} \tag{3}$$

where T_0 is the initial temperature, t_R – retention time, r – ramp (centigrade per time unit).

To illustrate this criterion, let us consider the boiling point of the series of thermally unstable alkyl azides, **R-N$_3$** in comparison with data for dialkyl diimides **R-N=N-R** (**Table 4**). In both series these data are available for homologs with $R \leq C_5H_{11}$. Boiling points of homologs with $R \geq C_6H_{13}$ at atmospheric pressure remain unknown at present due to instability of such compounds. Therefore, GC and/or GC–MS analysis of alkyl azides should be possible up to their retention temperature (T_R) not exceeding approx. $T_{lim} \approx$ 130–135°C (it has been confirmed experimentally [25]), and approx. $T_{lim} \approx$ 180°C for dialkyl diimides. This conclusion is confirmed by experimental RI values known just for some homologs of these series. On the other hand, within the alkyl hypochlorites series, **R-OCl**, T_b values are known only for members with R = CH_3 (9.2°C), R = C_2H_5 (27–36°C), and R = *tert*-C_4H_9 (77–78°C), meaning that T_{lim} for this series is not more than approx. 75–80°C. However, it is enough for separation of simplest homologs and determining their RI values, namely 502 (ethyl hypochlorite) and 605 (*tert*-butyl hypochlorite).

Within the series of aliphatic diazocarbonyl compounds with structural fragment **-CO-CHN$_2$**, ethyl diazoacetate, **N$_2$CHCO$_2$C$_2$H$_5$**, is one of the most often used reagents. Hence, the physicochemical properties available for this compound (including T_b 140–143°C) appear to be the most reliable comparing with data for other homologs [26]. We could not find in the literature the normal T_b values for higher alkyl diazoacetate homologs (with $R \geq C_2H_5$) just due to their instability, but we can conclude that GC and/or GC–MS analysis of other diazocarbonyl

R in R-N$_3$	T_b, °C	RI$_{non\text{-}polar}$	R in R-N=N-R	T_b, °C	RI$_{non\text{-}polar}$
CH_3	20–21	457	CH_3	1.5–2.0	396 ± 5
C_2H_5	48–50	543	C_2H_5	59	563 ± 4
C_3H_7	77.5–78	634	C_3H_7	113–115	756 ± 6
C_4H_9	106.5	746	C_4H_9	145.145.5	867 ± 2
C_5H_{11}	130–135	845	$(C_2H_5)_2CH$	182	1052 ± 16
$n \geq 6$	no data	no data	$n \geq 6$	no data	no data
T_{lim}	~ 130–135			~ 180	

Table 4.
Boiling points of some alkyl azides R-N$_3$ and dialkyl diimides R-N=N-R at atmospheric pressure and their GC retention indices on standard non-polar polydimethylsiloxane stationary phases.

compounds should be possible up to a temperatures of GC column approx. $T_{\text{lim}} \approx 140°C$. The applicability of this criterion has been verified during analysis of aryl substituted diazocarbonyl compounds [13]. For instance, such analyte as 1-diazo-4-phenylbuten-2-one, $\mathbf{C_6H_5CH_2CH_2COCHN_2}$, was analyzed under conditions ensuring
its retention temperature 145°C, slightly exceeding $T_{\text{lim}} \approx 140°C$. Exceeding the limit naturally leads to the appearance of a "plume" prior to the chromatographic peak of diazocompound that belongs to decomposition product, namely 4-phenyl-2-buten-1-one, $\mathbf{C_6H_5CH_2CH_2CH{=}C{=}O}$.

Thus, the sense of the chemical criterion for GC and/or GC–MS analysis of thermally unstable compounds is revealing their simple homologs for which the reference information on their normal boiling points (at atmospheric pressure) is available. The general recommendation to avoid the decomposition of unstable analytes in a GC column is not to exceed the column's temperature above this limiting value.

The criterion considered can be re-formulated, if necessary. If some homologs of any class of potentially unstable compounds indicate stability at sufficiently high retention temperatures, we can consider these T_R values as the limiting T_{lim} values for other homologs of the same series, or their structural analogues. Such viewing allows evaluating the really anomalous thermal stability of organic hydroperoxides, **R-OOH**, and, especially, peroxides, **RO-OR**.

Improvements of contemporary fused quartz capillary columns in GC (increasing of their inertness) permits us to use them, for example, for separation of obviously unstable hydroperoxides formed from monoterpene hydrocarbons in plant essential oils [27], cyclohexyl- and cycloheptylhydroperoxides and corresponding dicyclohexyl- (**I**) and dicycloheptylperoxides (**II**) [28]. Most exotic structures with peroxide fragments which can be separated using GC without decomposition are 3,3,6,6-tetramethyl-1,2,4,5-tetraoxocyclohexane (trivial name "diacetone diperoxide", **III**), 3,3,6,6,9,9-hexamethyl-1,2,4,5,7,8-hexaoxocyclononane ("triacetone triperoxide", **IV**), and even 3,3,6,6-tetrapropyl-1,2,4,5-tetraoxocyclohexane (**V**) [29]:

(III)

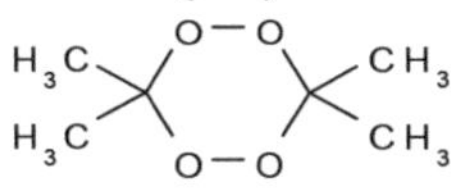

(IV)

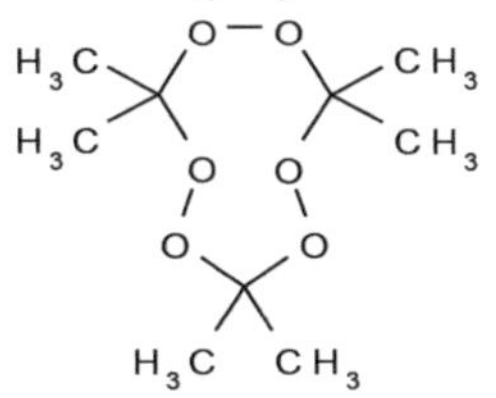

(V)

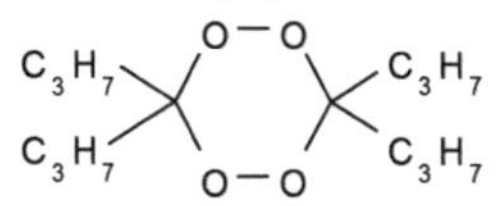

The most notable are the high retention temperatures of mentioned compounds: T_R value for dicyclohexylperoxide (**I**) is approx. 129°C, for dicycloheptylperoxide (**II**) ~ 180°C, for compound (**IV**) – 110-160°C (in different regimes), and for compound (**V**) – 185-260°C (!). This corresponds to the possibility of peroxides separation in almost any temperature regimes without risk of thermal decomposition of analytes and no needs for special control of separation conditions.

This conclusion was applied in the analysis of the unusual impurity found in the sample of benzyl alcohol $\mathbf{C_6H_5CH_2OH}$ [30] with the retention index 1894 ± 10 (semi-standard non-polar stationary phase RTX-5 MS) and the following mass spectrum, $m/z \geq 39$ ($I_{\text{rel}} \geq 2\%$), **M** is the symbol of molecular ions:

230(1)**M**, 213(2), 198(2), 197(8), 107(14), 105(6), 92(23), 91(100), 79(3), 77(6), 51(2), 39(2).

Attempts to identify this compound using the database [2] appeared to be unsuccessful. Nevertheless, the detailed interpretation of this mass spectrum together with GC retention index permits us to establish its structure unambiguously. The maximal signal with *m/z* 91 confirms the presence of benzyl fragment in the molecule, the peak with *m/z* 213 belongs to the ions [M – 17] ≡ $[M – OH]^+$, and the peak with *m/z* 197 – to the ions [M – 33] ≡ $[M – OOH]^+$. Combining the available chemical and spectral information, we can attribute solely the structure of dibenzyl ether hydroperoxide for this impurity [30]:

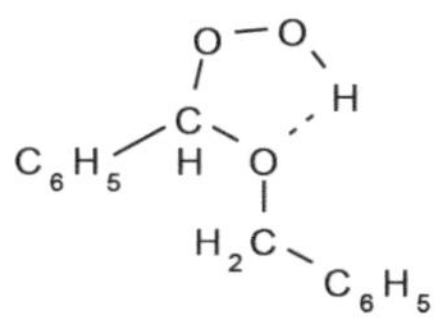

The existence and stability of such hydroperoxide seem rather unusual. At first, the presence of two functional groups (one of them with active hydrogen) at one carbon atom looks like very "exotic" structure of unstable hydroperoxide of semiacetal. The second feature is high retention temperature of this impurity (approx. 190°C). Nevertheless, in accordance with criterion mentioned above so high T_R value does not restrict GC separation of this compound without its thermal decomposition.

It is interesting to note that hydroperoxide of similar structure, $C_2H_5OCH(OOH)CH_3$, formed from diethyl ether, $(C_2H_5)_2O$, was detected by gas chromatographic analysis (RI 794) without its decomposition [31].

6. Conclusions

Two new examples of thermal decomposition of analytes during GC separation within a gas chromatographic column were revealed and considered. First of them is monomolecular decomposition of monoalkyl phthalates with the formation of phthalic anhydride. The second process – decomposition of non-substituted hydrazones of carbonyl compounds – seems like the first example of bimolecular reactions of analytes in a GC column. However, the visual manifestations of all these processes on the chromatograms are identical to the manifestations of interconversion of isomers, for example, keto-enol transformations of ethyl ester of acetoacetic acid in a GC column. The profiles of chromatograms in most cases contain a peak of initial analyte, a peak of a product of its decomposition or isomerization, and more or less expressed diffused "plateau" or "train" between them. Besides that, the first example of secondary chemical reaction in a GC column, when the decomposition by-product reacts with other constituents of samples, is revealed for hydrazones of alkyl aryl ketones.

Acknowledgements

Experimental data discussed in this work were determined by Lilia N. Fakhretdinova [20], Nikita E. Podol'ski [24], and Valentina M. Lukina [16] using the equipment of Resource Educational Centre "Chemistry" at the Institute for Chemistry of St. Petersburg State University. The author is grateful to the staff of this Center for assistance.

Author details

Igor G. Zenkevich
St. Petersburg State University, Institute for Chemistry, Universitetskii Prospect 26, St. Petersburg 198504, Russia

*Address all correspondence to: izenkevich@yandex.ru

References

[1] Ivanova N.T., Frangulyan L.A. GC analysis of unstable and reactive compounds. Khimia Publ., Moscow, 1979. (In Russian).

[2] The NIST 17 Mass Spectral Library (NIST17/2017/EPA/NIH). Software/Data Version (NIST17); NIST Standard Reference Database, Number 69, June 2017. National Institute of Standards and Technology, Gaithersburg, MD 20899: http://webbook.nist.gov (accessed: August 2020).

[3] Middleditch B.S. Analytical Artifacts: GC, MS, HPLC, TLC, and PC. J. Chromatogr. Library. Vol. 44. Elsevier, Amsterdam, 1989.

[4] Jain A., Hirata G.A., Farias M.H., Castillon F.F. Synthesis and characterization of (3-aminopropyl) trimethoxy-silane (APTMS) functionalized Gd2O3:Eu(3+) red phosphor with enhanced quantum yield // Nanotechnology. 27(6) (2016) 065601. doi: 10.1088/0957-4484/27/6/065601.

[5] Howarter J.A., Youngblood J.P. Optimization of silica silanization by 3-aminopropyltriethoxysilane // Langmuir. 22(26) (2006) 11142-11147. doi: 10.1021/la061240g.

[6] Zenkevich I.G., Ul'yanov A.V., Golub S.I., Buryak A.K. Chromatographic component of identification of the transformation products of 1,1-dimethylhydrazine in the presence of sulfur // Rus. J. General Chem. 84(6) (2014) 1102-1110. doi: 10.1134/S1070363214060097.

[7] Gough, T.A.; Webb, K.S. The trace detection of some non-volatile nitrosamines by combined gas chromatography and mass spectrometry // J. Chromatogr. 95 (1974) 59-63. doi: 10.1016/s0021-9673(00)84095-9.

[8] Zenkevich I.G., Yandovskii V.N. Chromato mass-spectrometric characterization of substituted 4-acyl-1,3,4-oxadiazolines // Chem. Heterocycl. Comp. 5 (1984) 498-505.

[9] Ioffe B.V., Kuznetsov M.A., Chernyshov V.A., Kuznetsova L.M., Zenkevich I.G. Heminal azohydrazines – the new class of organic compounds // Rus. J. Org. Chem. 12(11) (1976) 2273-2281. (In Russian).

[10] The Sadtler Standard Gas Chromatography Retention Index Library. Vols. 1-4. Sadtler Division, Bio-Rad Laboratories, Philadelphia, PA, 1986.

[11] Zenkevich I.G. Application of Recurrent Relationships in Chromatography // J. Chemometrics 23 (2009) 179-187. doi: 10.1002/cem.1214.

[12] Zenkevich I.G., Bystrova G.I. Quantitative analysis of mixtures unstable at the conditions of chromatographic separation. Determination of propylene glycol impurity in propylene carbonate // J. Analyt. Chem. (Rus.). 40(9) (1985) 1686-1693.

[13] Kornilova T.A., Ukolov A.I., Kostikov R.R., Zenkevich I.G. A simple criterion for gas chromatography/mass spectrometric analysis of thermally unstable compounds, and reassessment of the by-products of alkyl diazoacetate synthesis // Rapid Commun. Mass Spectrom. 27(3) (2013) 461-466. doi: 10.1002/rcm.6457.

[14] Masur M., Grutzmacher H.-F., Munster H., Budzikiewicz H. Mass spectrometric fragmentation of the tautomers of 1,3-diketones // Org. Mass Spectrom. 22 (1987) 493-500.

[15] Allegretti P.E., Schiavoni M.M., Di Loreto H.E., Furlong J.J.P., Della Vedova C.O. Separation of keto-enol tautomers in β-ketoesters: a gas

chromatography – mass spectrometric study // J. Mol. Struct. 560 (2001) 327-335.

[16] Zenkevich I.G., Lukina V.M. Characteristic features of the gas chromatographic separation of tautomers of ethyl acetoacetate // J. Phys. Chem. A. 2020 94(6) 1214-1223. doi: 10.1134/S0036024420060357.

[17] Trapp O., Shelle R., Marriott P., Schurig V. Simulation of elution profiles for two-dimensional dynamic gas chromatographic experiments // Anal. Chem. 75(17) (2003) 4452-4461. doi: 10.1021/ac0301144.

[18] Practical Gas Chromatography. Eds. W. Engewald, K. Dettmer-Wilde. Springer-Verlag. Berlin, 2014. 902 p.

[19] Shinde Y., Sproules S., Kathawate L., Pal S., Konkimalla B., Salunke-Gawali S. Separation and isolation of tautomers of 2-hydroxy-4-naphthoquinone-1-oxime derivatives by liquid chromatography: antiproliferative activity and DFT studies // J. Chem. Sci. 2014 126(1) 213-225.

[20] Zenkevich I.G., Fakhretdinova L.N. Thermal instability of monoalkyl esters of phthalic acid during their gas chromatographic separation // Analytics & Control. 19(2) (2015) 175-182. doi: 10.15826/analitika.2015.19.2.013. (In Russian).

[21] Khaibulova T.S., Boyarskaya I.A., Larionov E., Boyarskiy V.P. Cobalt-catalysed methoxycarbonylation of substituted dichlorobenzenes as an example of a facile radical anion nucleophylic substitution in chloroarenes // Molecules, 19 (2014) 5876-5897. doi: 10.3390/molecules19055876.

[22] Hamming M.C., Foster N.G. Interpretation of mass spectra of organic compounds. Academic Press. New Yotk. 1979. 694 p.

[23] Kuethe J.T., Childers K.G., Peng Z., Journet M., Humphrey G.R., Vickery T., Bachert D., Lam T.T. A practical kilogram-scale implementation of the Wolff-Kishner reduction // Org. Process Res. Development. 13(3) (2009) 576-580. doi: 10.1021/op9000274.

[24] Zenkevich I.G., Podol'skii N.E. Revealing compounds unstable during gas chromatographic separation. Non-substituted hydrazones of carbonyl compounds // Analytics & Control. 21(2) (2017) 125-134. doi: 10.15826/analitika.2017.21.2.002. (In Russian).

[25] Akhlaghinia B., Samiei S. A novel and highly selective conversion of alcohols, thiols, and silyl ethers to azides using the 2,4,6-trichloro[1,3,5]triazine / n-Bu4NN3 system // J. Braz. Chem. 18(7) (2007) 1311-1315. doi: 10.1590/S0103-50532007000700003.

[26] Clark J.D., Shah A.S., Peterson J.C. Understanding the large-scale chemistry of ethyl diazoacetate via reaction calorimetry // Thermochimica Acta. 392-393 (2002) 177-186. doi: 10.1016/S0040-6031(02)00100-4.

[27] Rudback J., Ramzy A., Karlberg A.-T., Nilsson U. Determination of allergenic hydroperoxides in essential oils using gas chromatography with electron ionization mass spectrometry // J. Sep. Sci. 37(8) (2014) 982-989. doi: 10.1002/issc.201300843.

[28] Wierzchowski P.T., Zatorski L.W. Determination of cyclo C_6 and C_7 peroxides and hydroperoxides by gas chromatography // Chromatographia. 51(1-2) (2000) 83-86.

[29] Barreto G.P., Canizo A.I., Eyler G.N. Use of retention data as the first step in the identification of cyclic organic peroxides in temperature-programmed gas chromatography // Chromatographia. 63(5-6) (2006) 261-266. doi: 10.1365/s10337-006-0716-y.

[30] Zenkevich I.G. Special features of gas chromatographic determination

of dibenzyl ether hydroperoxide admixture in benzyl alcohol // Rus. J. General Chem. 86(9) (2016) 2016-2021. doi: 10.1134/S1070363216090061.

[31] Delort E., Jaquier A. Novel terpenyl esters from Australian finger lime (*Citrus australasica*) peel extract // Flav. Fragr. J. 24(3) (2009). doi: 10.1002/ffj.1922.

Chapter 4

Recent Advances in Targeting Clinical Volatile Organic Compounds (VOC)

Imadeddine Azzouz, Mohammad Sharif Khan, Andrew C. Bishop and Khaldoun Bachari

Abstract

This chapter introduces the significance of exploring volatile organic compounds (VOC) in clinical samples. Because exhaled-breath is easy to collect, unlimited, and instruments are already commercially available, VOC analysis in exhaled breath seems to be a promising tool for non-invasive detection of many diseases including infections, respiratory diseases, and cancers. Here, we have focused on some appropriate technologies to extract, pre-concentrate, and evaluate VOC biomarkers in exhaled breath. The second part of this chapter discusses the comprehensive GC × GC in bio-VOCs analysis and illustrates the potential of using this analytical technique.

Keywords: gas chromatography, breather analysis, Sampling, Pre-concentration, VOCs, Biomarker, GC–MS, GCxGC, Mass spectrometry

1. Introduction and scope of the chapter

Inside the human body, cells produce hundreds of biochemical reactions at extremely precise and controlled moments. Cells can be thought of as a factory, regulating what enters or leaves its barrier. From a metabolism point of view, any "error" in biochemical processes (temporary or permanent) leads to abnormal concentrations of metabolites and/or presence of "abnormal" metabolites. Consequently, a wealth of metabolites with low molecular weight as well as high molecular weight can be exploited as a "precious" source of information revealing the metabolic state of the body. Metabolites can be excreted via the urine, feces, saliva, blood, breath, or sweat. Among these metabolites, researchers are actively trying to find biomarkers, identifying the presence of different diseases (cancers, infections and so on). Especially in the instance of different cancers, the lack of specific syndromes with limited understanding of etiology make it difficult to diagnose at an early stage. Nevertheless, biomarkers can turn out to be powerful tool in predicting the development of these cancer and other diseases.

Among all biological samples, exhaled breath has many advantages compared to bio-fluids. First, breath sampling is pain-free, non-invasive, and most important is almost "unlimited". Secondly, breath Volatile Organic Compounds (VOC) are collected from the airways which is directly connected to the entire body via the bloodstream. Blood continuously circulates around body periodically reaching the air-blood barrier of the alveoli within lungs, VOCs from the whole body can cross this barrier from the blood and be released into the exhaled breath. Conversely, exhaled breath is a very complex matrix and can be challenging to investigate being influenced by a patient's habits, diet, and the environment [1].

The human airways emits a gigantic number (>3000) of VOCs of different origin [2]. VOCs can either be endogenous (arise from the body) or exogenous (environmental source) as shown in **Figure 1(A)**. While it is rather easy to monitor exogenous VOCs, endogenous VOCs can be produced from various sources: normal metabolism of nutrient, inflammatory processes, metabolic processes (diseased and normal), cancerous cells, and microbiome of the oral cavity, airways and gastrointestinal tract. In addition to multiplicity of compounds originating from the different endogenous sources, exhaled breath is saturated with water vapor leading to relative humidity close to 100%, which may impact the collection and analysis of VOCs. In spite of these challenges in exhaled breath research, it continues to attract the interest of scientists worldwide. For the past fifty years, publications on VOCs and exhaled breath have grown exponentially with great effort being devoted to discover VOCs biomarkers related different diseases (**Figure 1(B)**).

A typical workflow for breath collection and analysis in a clinical setting is shown in **Figure 2**. The breath analysis starts with the study design and sample

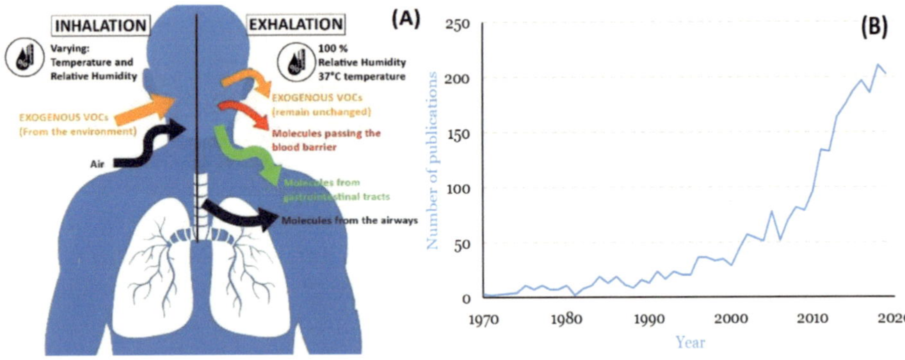

Figure 1.
(A) Pathway of exhaled molecules in the human body, (B) number of research publications involving VOCs and breath field (data from PubMed database).

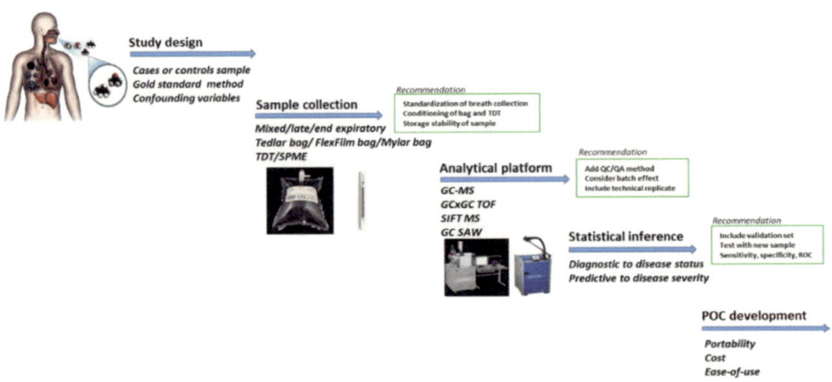

Figure 2.
A schematic of breath analysis pipeline and recommendation. TDT: Thermal desorption tubes; SPME: Solid phase microextraction; SIFT MS: Selected-ion flow-tube mass spectrometry; GC- MS: Gas chromatography–mass spectrometry; GCxGC TOF: Two-dimensional gas chromatography and time-of-flight mass spectrometry; GC SAW: Gas chromatography surface acoustic wave.

collection. Once collected samples must be properly transported and stored until analysis. After the analysis on an analytical instrument important for the specific study design, data would undergo statistical testing and biological interpretation. Finally, results require validation before being directed to the Point-of-Care (POC) development stage.

This book chapter addresses the efforts of many researchers to identify and validate VOCs resulting from various diseases and summarize important technological advancements used to pre-concentrate and analyze VOCs.

2. Capturing, focusing, and storing breath samples

The concept of "off-line" breath analysis can be broadly fragmented into breath sample collection, sample analysis, and data analysis. In this section, a summary of principal methods used to "catch" and to "focus" breath samples is exposed whilst advantages, disadvantages, and practical suggestions are systemically reported.

2.1 Sampling

The major objective of sampling is to take a representative "sample" from a matrix. Usually, a pump or vacuum are used to achieve this process. Decisions to be made must balance between cost including the duration of the sampling period, the size of each sample, and the number of samples.

2.1.1 Bags

Sampling bags are low-cost, whole-air sampling devices for VOCs and permanent gases. Several EPA, NIOSH, and OSHA methods exist for bag sampling for a variety of applications: sources emissions; indoor air quality, workplace atmospheres, and breath analysis [3]. Bags remain popular among researchers, principally due to their low cost and reusability. However, they are known to contain several artifacts, and a tradeoff between competitive pricing and performance is rapidly pointed.

Tedlar® bags are the most commonly used polymer-based bags in air and breath research. Their main disadvantages to the use of Tedlar® bags are largely due to the interaction of air or breath constituents with the polymer which bags are fabricated with. This interaction results in contamination through emission, adsorption, and diffusion. Other polymers were developed to overcome the above-cited disadvantages, such as Mylar (polyethylene terephthalate) and Altef (polyvinylidene difluoride). **Table 1** presents a brief comparison of different type of sampling bags.

Some practical considerations must be taken to avoid sample loos or degradation. Moreover, it is preferable to not reuse bags.

Practical considerations:

- Clean bags before use to minimize the background signal associated with the plastic interior (several times).

- Bag cleaning should be performed as close to the time of sampling (flushing nitrogen for example).

- Analyze or pre-concentrate samples already filled in the bag as fast as possible to avoid losses, interaction with the plastic of the bag, photodegradation, and adsorption.

	Tedlar bags	Multi-layer foil bags	Altef bags
Composition	PVF	4 layers: Nylon, 2 × polyethylene, aluminum foil	PVDF
Advantages	• EPA testing methods • Bag available with stainless steel valves	• Suitable for H_2S • Very low permeability to O_2 and CO_2	• Low VOC background • Lower permeability than Tedlar toward CO_2 and N_2
Limitations	• Background level of phenol and DMAC • High permeation of CO_2 and O_2	• Should analyze within 48 hours especially for CH_4, H_2S, CO, and CO_2	• Less resistance to UV light than Tedlar • Should analyze within 24 hours

Table 1.
DMAC: Dimethylacetamide, PVF: Polyvinyl fluoride, PVDF: Polyvinylidene fluoride.

- Protect the bag from direct sunlight and store it in a rigid container to prevent photodegradation and bag puncture respectively.
- Do not fill the bag more than 80% of its volume

2.1.2 Canisters

Despite their price that is expensive compared to sampling bags, canisters are known to be robust, relatively inert, and non-permeable (example of breath air [4, 5]). They are made with stainless steel and their inner surface interaction with the samples (adsorption, desorption) is of paramount importance especially when lower and lower concentrations exist. The chemical composition of the metal obviously will affect the type of chemical or physical reactions with the sample. To enhance their inertness vis-à-vis sulfur-based samples (for example), canisters are passivated, electropolished, or coated. Canisters are known to be reusable and the sample can be stored until 30 days without loss or degradation.

Practical considerations:

- Canisters are less useful, is some cases, to the storage of semi-volatile and polar compounds DUE to condensation and/or dissolution into water at higher pressure.
- Choose inner surface-deactivated to avoid adsorption and interaction with samples.
- Canisters must be cleaned prior to use. In fact, canisters should be pressurized and evacuated (to be cleaned), and evacuated once more to create vacuum.

An alternative to metal canisters, glass containers have been used for sampling breath [6]. Despite their fragility, they can be more performant than Tedlar bags as reporter by Scott-Thomas et al. [7].

2.1.3 Other containers

Other breath collection container were reported including: gas-tight syringes [8], face mask [9], glass tube [10], and gas bulb [11]. To go further, Lawal et al. [12] investigated breath sampling methods by performing an in depth bibliometric search.

2.2 Focusing

Direct sample introduction by syringe or rotary valves is only suitable for small volumes of "relatively" concentrated samples. In the field of metabolomics, the discovery of diseases biomarkers for example (from urine, feces, breath, tumor and cells) requires analysis of trace and ultra-trace levels of targeted compounds.

Traces analysis methods involve analyte accumulation by sorbent (solid, film) followed by thermal vaporization in the presence of a flow of gas to transport them to the analyzers.

2.2.1 Thermal desorption

Thermal desorption tubes (TD) are the most commonly used medium for the collection and pre-concentration of human breath samples for cancer diagnosis, infections, and bacteria recognition [13, 14]. In thermal desorption technique, sample is swept into the gas chromatograph using heating and a flow of the carrier gas. The desorbed "plug" or "band" of the sample should be as a narrow as possible (chromatographic considerations). However, due to low mass transfer, commonly, sample is first heated slowly letting the desorbed material to be cold-trapped at the head of the column (cold trap or a cryogenic oven). Secondly, the re-condensed sample is then desorbed as the temperature program proceeds.

To trap the VOCs, thermal desorption tubes comprise various sorbent materials (**Table 2**). The sorbent materials can be synthesized (polymer such as Tenax TA) or obtained by graphitizing carbon, which adsorbs a large molecular range of VOCs.

The small size of the tubes (~7 cm length), and their suitability to be used both for active and passive sampling (with or without pump) make them attractive for various applications even standardized methods (EPA—TO-17, ASTM—D6196, NIOSH—2549).

Supelco reported a tool for selecting adsorbent for thermal desorption applications [15]. The goal is to select the "proper" adsorbent that can retains a specific or groups of analytes for a specified sample volume.

Sorbent name	Material	Applications
Tenax TA	2,6-diphenylene-oxide porous polymer	C7-C26
Tenax GR	Mixture between graphite (30%) and (70%) 2,6-diphenylene-oxide	C7-C26
Carbotrap	Graphitized carbon black	
• X		C3-C9
• B		C5-C12
• C		C12-C20
• F		>C20
Carboxen	Carbon molecular sieve	
• 1016		C3-C5
• 10xx		C2-C5
Carbosieve	Carbon molecular sieve	C2-C5
• G, S-II, S-III		

C: Number of carbon atoms.

Table 2.
Various sorbent materials used for thermal desorption applications.

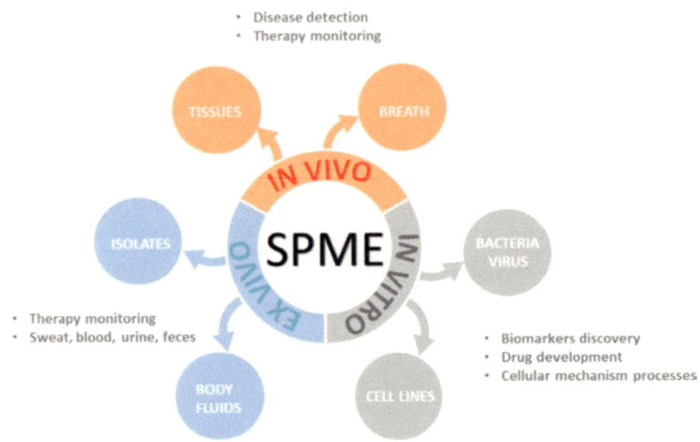

Figure 3.
Common applications of SPM in biology and medical researches.

2.2.2 Solid-phase micro extraction (SPME)

Solid-phase Microextraction (SPME) technic uses a polymer-coated fiber housed in a modified syringe as a sampling device. When SPME is used for analysis, first, the syringe needle is placed into the analyte, and the coated fiber is then exposed. Once the system is brought to equilibrium, the coated fiber is retracted into the syringe needle and removed from the sample (bag filled with breath for example). The needle is then transferred to a heated inlet of the GC, and the analytes are thermally desorbed.

Although SPME is well established, accepted, and validated for various fields (air quality, environment, and food analysis), having yet to gain acceptance as a standard method in biotechnological industries. **Figure 3** highlights exemplary applications of SPME toward health monitoring and biomedical research.

The volatilome of the healthy human body comprises over 1840 VOCs (breath, blood, sweat, urine, and feces) [16]. In this line, SPME-GC–MS was used by Garcia et al. [17] to compare breath issuing from smokers, non-smokers, and patients with laryngeal cancer. Authors have found seven unique VOCs discriminating non-healthy and healthy controls. In the same line, SPME was applied in vitro to demonstrate that particular VOCs are present in exhaled breath of lung cancer patients at significantly different levels than those found in healthy controls [11].

Bean et al. [18], with the help of SPME-GCxGC-TOF-MS, identified 70 compounds indicating presence of *Pseudomonas aeruginosa* bacteria. Similarly, SPME-GC–MS was complemented with SIFT-MS to differentiate 15 clinical strains from 5 environmental strains of *Stenotrophomonas maltophilia* (responsible of lung infection) [19].

Type of coating	Film thickness (μm)	Polarity	Maximum temperature (°C)	Core type
PDMS	30, 100	Non-polar	300	Fused Silica
				Metal
PA	85	Polar	320	Fused Silica
PEG (Carbowax)	60	Polar	250	Metal
Carbopack Z/PDMS	15	Bipolar	340	Metal
PDMS/DVB	65	Bipolar	270	Stableflex
				Metal
Carboxen-PDMS	85	Bipolar	320	Stableflex
				Metal

PDMS: Polydimethylsiloxane, DVB: Divinylbenzene, PA: Polyacrylate, PEG: Polyethylene glycol.

Table 3.
Type of commercially available SPME fiber adapted from Ref. [20].

Type of SPME fiber significantly affect the number and the type of volatile compounds that can be detected, and hence introduces another source of variability to the results [17]. Common type of SPME fibers used to extract compounds are listed in **Table 3**.

PDMS is the coating of choice to extract many classes of non-polar and less polar compounds. At the opposite, polyacrylate coating is the best to extract polar compounds. For example, mixed phase coatings (polar and non-polar) based on PDMS-PEG are used to extract both polar and non-polar compounds.

In addition to the above-cited applications of SPME fibers, their ability to extract semi-volatile or non-volatile compounds was demonstrated [21, 22]. In fact, biofluids including urine, saliva or blood, might comprise many organic compounds which are not present in the vapor phase and hence could not been achieved by headspace analysis [23, 24]. For additional information, a recent review compiling the most recent applications of SPME in biotechnology and clinical studies was written by Filipiak et al. [25].

2.2.3 Other technics

Headspace takes advantage of the closed-vessel equilibrium between either a solid or a liquid and a gas (urine, blood, feces). In the headspace analysis, an aliquot of the equilibrated gas phase is removed from the vessel and GC analyzes the aliquot. This technic is often coupled to SPME (exposing the fiber to the headspace of the sample). Zhang and Raftery reported metabolic profiling of urinary VOCs using SPME-GC–MS [26].

2.3 Sampling breath portions

VOCs present in exhaled breath may also originate from external environment (exogenous VOCs). The goal of sampling breath is to minimize the concentration of exogenous substances. Additionally, dead space air (oral and nasal cavities, gut) acts to dilute and may contaminate VOCs from blood gas exchange. To subtract VOCs levels from contaminant sources, measurement follow specific pathway. Using a capnometer, Miekich et al. [27] found that the end-tidal portion of breath showed the highest concentrations of endogenous and the lowest concentration of exogenous substances. In addition to end-tidal, other terms as Mixed respiratory or late expiratory are found to describe the rest of breath fractions [28].

2.4 Breath sampling apparatus

2.4.1 R-tube

R-tube is well known as exhaled breath condensate portable collector. The sampler includes a cooling sleeve (frozen until needed), a one-way valve, and a plunger. A modified version or R-tube was used by Martin et al. [29] to sample breath volatile compounds by SPME.

2.4.2 Bio-VOC

Bio-VOC is a commercially available system marketed by Markes International. It consists of a hard plastic cylinder connected on one side to a disposable mouthpiece and to sorbent tube on the other side. Poli et al. [30] have used Bio-VOC with SPME fiber to determine aldehydes in exhaled breath of patients with lung cancer.

2.4.3 ReCIVA sampler

Reciva sampler (marketed by Owlstone Medical) is a commercially available sampler. Its uses pressure-modulation (two pumps) breath sampling onto four adsorbent tube. A disposable silicon mask coupled to bacterial filter are also present to prevent cross-contamination (contamination between subjects).

A pressurized clean air supply is however needed to minimize the concentration of exogenous VOCs.

3. Online and offline breath analysis (versus or collaborative)

"On-line analyses" of exhaled breath, sometimes called "direct" or "real-time" breath analyses are mass spectrometry based methods which needs no sample preparation or collection. At the opposite to off-line methods which requires two-step processes (collection and analysis), on-line analyses are one-pot analysis method. The three main online methods for breath analyses are SIFT-MS, PTR-MS, and SESI-MS.

SIFT-MS is more suitable for targeted analyses however their limiting factor is low mass resolution. Recently, new enhancement is introduced to this method including the use of time-of-flight (TOF) mass analyzers and the use of the electric fields in the drift tube (SIFDT).

PTR-MS had been first established in the field of environmental analyses and recently in the field of breath analyses. Its limitation is set by the principle that only VOCs having a higher proton affinity than that of water can be detected. New enhancement including the use of TOF and more recently a quadrupole boosts this method to detect higher than m/Δm 10 000.

SESI-MS has been found to be the most sensitive for polar compounds reaching sub ppqv. The use of SESI-MS in analyses of the breath condensate is well established by high-performance gas chromatography (HPLC). However, the main disadvantage of this method is that use to analyze gas phase is not yet possible.

Undoubtedly, Mass Spectrometric methods are the most sensitive and the most appropriate for compounds discovery and identification. However, their use in clinic studies remain limited due to practical considerations. Bruderer et al. [31] reviewed extensively on-line methods for exhaled breath analyses.

4. Two-dimensional GC

In the early 90's, two-dimensional gas chromatography has become available for the separation and the identification of complex mixtures. This section demonstrates the versatility and applicability of 2D-GC and an analytical tool for breath analyses.

4.1 Principle

In GC, higher the resolving power, the better the performance. It is understandable that, if the column becomes longer, they could resolve more analyte. There are multiple examples of a very long (>100 m) column used on a very complex sample separation in a 1D GC experiment [32]. Other than the length of the column, there are few more approaches have been taken to improve the resolving power of the samples. One, forwarding a part of the sample to a second column to be better resolved based on another separation mechanism. This process is known as the Classical Multidimensional Gas Chromatography or MDGC [33]. The second approach transfers effluent of the first column to a second column for a comprehensive separation of the samples by a modulation device that forwards a very narrow band from the first to the second column. This process is named as GC × GC or Comprehensive two-dimensional gas chromatography [34].

The GC × GC system is operated as two different stationary phases that have different separation mechanisms are connected to each other so result in an orthogonal separation of the sample [34]. But to achieve this, there is a special device called "modulator" is connected in between the two phases, so it can sample and re-inject the first column effluent to the second phase for a complete independent separation [35].

The heart of GC × GC is the modulator. There are extensive reviews that already exist on the fundamental of the GC × GC modulator [36, 37]. There are different types of modulators that have been introduced in the field, such as thermal modulator, valve-based modulator and, flow modulator. The main task of all these modulators are to sample the primary column effluents to the second column (a sharp pulse of 1D peak). One main difference is the requirement of the focusing steps. The cryogen-free modulator such as the flow modulator, Peltier modulator does not have any focusing effect where the cryogenic modulator like longitudinally modulated cryogenic system (LMCS) has focusing steps, hence requires liquid N_2 or CO_2 to cold trap the analyte from 1D before eluting to the 2D column. Although it is a compensation of the peak capacity, the consumable-free modulator is getting more popular due to low set-up fees, less consumable and maintenance free and lowest safety issue. **Figure 4** gives an example of 2D-GC configuration.

This is vital for the detector of GC × GC to have certain characteristics. This is because the second dimension separation occurs quickly, demanding a fast acquisition detector with at-least 50 scans/second. The Flame Ionization Detector (FID), Electron capture detector (ECD) or Flame photometric detector (FPD) also acquire fast data and are able to collect enough data points for even 300 ms width GC × GC peak. The high-speed MS detector time of flight (TOF) is often used as a detector for the GC × GC. Using MS with GC × GC separation

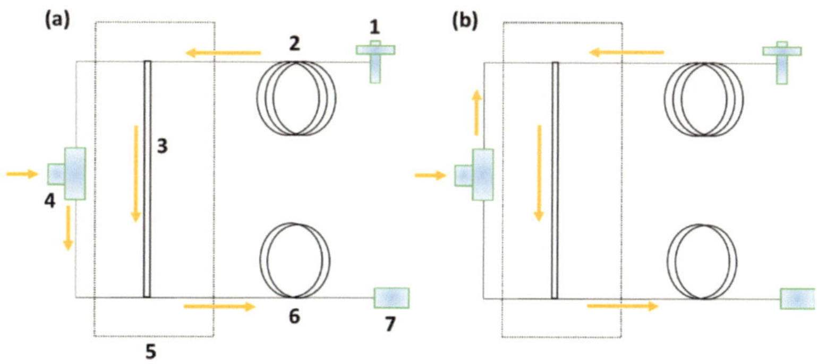

Figure 4.
The capillary flow technology (CFT) based flow modulator design. The arrows show the flow direction. The injector (1) is connected to the primary column (2) that passes through a channel (3) that is mounted inside of CFT device (5). A modulation valve (4) is connected to supply ancillary flow of carrier gas supply to divert the flow in a downward direction as seen in (a): Load position, or upward as in (b): Inject position toward to a 2D column (6) to detector (7). With permission from [35].

adds massive benefits for compound identification through the deconvolution of the mass/charge ratio (m/z) and results presented in a two dimensional contour plot. There are few software package available to do this deconvolution and identification of compounds with spectral matching to known compound libraries.

The primary benefit of using GC × GC for VOCs analysis is the enhanced selectivity that helps complete separation of the complex VOCs matrix. A number of literature has been reported to utilize the GC × GC for this purpose, and discussed more detail in the next section. The typical column configuration that are used with GC × GC for the breath analysis are a non-polar long column (30 to 60 meters) such as (5%-phenyl)-methylpolysiloxane phase as first dimension and a short (1–2 meters) polar column such as Wax phase as second dimension [13, 14, 18]. An interesting optimization study also recommend the same non-polar and polar column set for VOCs analysis [38]. The study compares a number of commercially available column such as 1%-phenyl, 5%-phenyl, and WAX phase as ^{1}D column and WAX, 5%-phenyl and ionic liquid column as ^{2}D phase. Based on the observed chromatogram and tradeoff between column maintenance and separation efficiency, the orthogonal non-polar and polar set was recommended.

Nevertheless, the 1D GC is long been used for the VOCs analysis as they are the most robust and reliable tools for the VOC analysis and available in many labs around the world. The benefits of using 1D classical GC for VOCs include the reproducibility, simple operation, presence of the comprehensive library and ease of data management. There are a number of mass analyzer technologies such as quadrupole, TOF and Ion trap that also makes 1D GC as a strong tool for the VOCs analysis. The 1D GC–MS systems is reported to analyze VOC "propofol", a drug that is used as anesthesia during surgery [39]. These results indicated a precise and sensitive determination of propofol in breath and blood by the GC–MS analysis. Ulanowska et al., determined the VOCs from the *Helicobacter pylori* as an indicator of gastrointestinal infection [40]. Their results identified a panel of breath metabolites, isobutane, 2-butanone and ethyl acetate as potential biomarker of the *H. pylori* infection.

4.2 Applications

Breath analysis is traditionally used for the signature of the disease. There is a disease-specific chemical signature that can be differentiated based on the breath

Common breath biomarkers between Human and Monkey (total match is 40.8%)		
MW (g/mol)	Human	Monkey
	Exact match (10.5%)	
114.2	Heptanal	Heptanal*
170.3	Dodecane	Dodecane
184.4	Tridecane	Tridecane
168.3	hexylcyclohexane	Hexylcyclohexane
	Isomer match (30.3%)	
136.2	l-beta-pinene	Trans-á-Ocimene
	camphene	beta-phellandrene
142.3	Heptane, 3-ethyl-2-methyl-	2,3,6-trimethylheptane
	Octane, 2,6-dimethyl-	-
	5-ethyl-2-methylheptane	-
170.3	3,7-dimethyl-decane	2,5-dimethyldecane
170.3	4-ethyl-5-methyl-nonane	2,2,4,6,6-pentamethylheptane
184.4	4-methyl-dodecane	4-ethyl-2,2,6,6-tetramethylheptane
120.2	1,3,5-trimethyl-benzene	1,2,3-trimethylbenzene
226.4	Pentadecane, 5-methyl-	Hexadecane
128.3	Heptane, 2,4-dimethyl-	2,2,3-trimethylhexane
	Nonane	-
	2,3-dimethyl-heptane	-
168.3	4,6,8-trimethyl-1-nonene	-

* confirmed by authentic standard

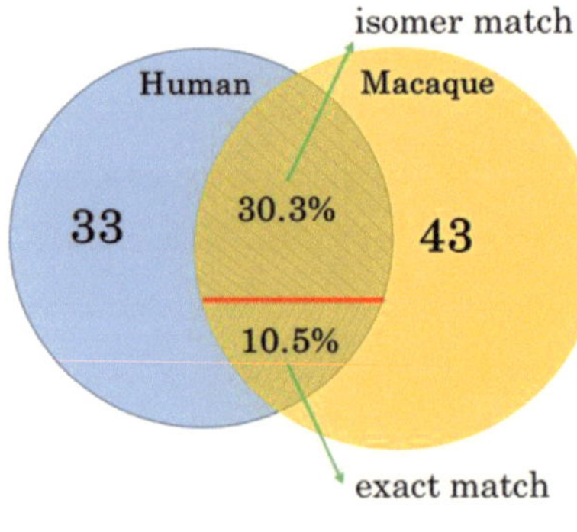

Figure 5.
The VOCs from the NHP and humane identified by similar GC × GC TOF MS. The VOCs signature compounds from the different pathway (prepared from the data published on [14, 39]).

sample. Numerous example has been published where the breath samples were analyzed and compared with the healthy person breath to determine the VOCs that express differently from the diseased person. A good example is Tuberculosis, this is a disease caused by a bacteria called "*Mycobacterium tuberculosis*". The main residing location of these bacteria is in the lung when they interact with the lung cell and use the energy pathway to proliferate. Hill group demonstrates the VOCs from the culture of *Mtb* are different than the VOCs from the empty culture using the GC × GC TOF instrument [41]. Using a similar GC × GC method, they have also found a difference between Mtb infected Non-Human Primate (NHP) breath from uninfected breath [13, 42, 43]. The match between the documented VOCs is listed below. The heptanal, dodecane, tridecane, and hexylcyclohexane were detected from both NHP and humane infected breath VOCs. **Figure 5** shows the match between the VOCs detected from human and NHP. Using a similar instrument platform, almost 50% of VOCs could be matched between the two species' breath.

In addition to breath, GC × GC TOF MS has become an essential tool for metabolomics. This process tries to look at the volatile, semi-volatile, and heavy boiler compounds from the biological fluid such as serum or plasma, cell, urine, and other biofluids [44]. Mishra et al. reported an interesting comparison study between the GC × GC and 1D GC for the serum sample. According to the author, the GC × GC can detect about 5,000 metabolites whereas the 1D GC could only detect about 500 metabolites. It might seems overwhelming as the 1D GC was equipped with high-resolution Orbitrap MS which is high resolution but less sensitive than the QTOF used with the GC × GC system [45]. Yu et al. optimized the GC × GC parameters to identify the maximum number of metabolites including the VOCs from the biofluids [46]. Urinary aromatic amine is an indicator of cigarette smoking. SPME of urine volatiles of smokers and analysis by GC × GC MS has revealed more than 150 aromatic amine compounds which is a much higher number from the previous 1D GC analysis [47]. VOCs analysis of biological fluid using the GC × GC techniques is still raising and within the next couple of years, this process will become a more prominent and established process for the VOC analysis.

4.3 Comparison with other analytical methods

There are several methods currently utilized by dozens of research groups to analyses the breath VOCs. The essential question here to ask which type of method might be a stronger candidate for the breath volatiles. There are many well-documented reviews summarized the analytical methods used for the breath VOCs analysis [48–53]. Essentially the basic method for analysis breath biomarkers is classified based on the objective of the study. For instance, the chromatography separation and mass spectroscopy for the purpose of identification (untargeted) or quantification (targeted) analysis. A comparison table between the instrumental platform used for the VOCs analysis is provided in **Table 4** with the advantages and disadvantages of each method. The critical comparison of these techniques is not intended to describe here but it can easily see the high solving power of the GC × GC method put it as the highest sensitive device currently available for the VOCs analysis. To sum up this discussion, all analytical platforms used for VOCs analysis could be classified into few subcategories (a) sensor array for detecting any specific analytes (b) separating a mixture of analytes from the matrix, and (c) the high-resolution mass spectroscopy for identification. Reader are referred to Sethi et al. [54] for more detail comparison.

Intrumental platform	Advantages	Disadvantages
Comprehensive two dimentional GC × GC	High sensetive identification Reproducible quatification Maximum number of VOCs	Required special tools and software
Gas chromatography with mass spectroscopy	Sensitive at ppb level Useful to identify unknown Robust and reproducible	High temperature operation Known standard requires Not cost effective
Ion mobility spectrometry	Sensitive at ppm level Portable Cost effective	Chemical fingerprint and identification are not possible Offline analysis Low sensitive
Selected ion flow tube mass spectrometry	Sensitive at ppb level Rapid and cost effective Portable and online	Chemical fingerprint and identification are not possible
Proton transfer spectrometry	Sensitive at ppb level Direct injection of VOCs Online measurement and monitoring	VOC chemical identification and complete profiling not possible
Various chemical sensor matrix platforms/e-noses	Sensitive at ppb level Cost effective and portable Fast and easy Point of care compatible	VOC chemical identification and complete profiling not possible

Table 4.
Advantages and disadvantages of some of the methods currently used for VOC analysis in clinical matrices.

5. Conclusion

Clinical VOCs may have potential for noninvasive pathological diagnosis. However, discovering biomarkers associated with specific disease requires standardization methods for both sampling and analyzing. The present contribution discusses recent advances in analyses of exhaled breath VOCs and focuses on chromatographic technics for off-line analyses. Mass spectrometric technics for on-line analyses were illustrated and their potential for VOCs discovering were demonstrated. The last part of this chapter discussed the comprehensive GC × GC technic and its ability in bio-VOCs analyzing. This technic has been proven as effective due to enhanced peak capacity and sensitivity.

There are several research centers working on the two competitive technics (on-line and off-line) and significant amount of resources are dedicated. We believe that breath analyses will become the method of choice for diagnosis of several pathologies such as Asthma and other infections. As discussed above, standardization is crucial to go further in the process of validation with a sufficient cohort size.

Author details

Imadeddine Azzouz[1,2*], Mohammad Sharif Khan[3], Andrew C. Bishop[3] and Khaldoun Bachari[1]

1 CRAPC Centre de Recherche Scientifique et Technique en Analyses Physico-Chimiques, Tipaza, Algeria

2 ESIEE Paris, University Gustave Eiffel, Noisy-le-Grand, France

3 Center for Precision Medicine, Department of Internal Medicine, Section on Molecular Medicine, Wake Forest School of Medicine, Nutrition Research Center, Winston-Salem, NC, USA

*Address all correspondence to: imadeddine.azzouz@esiee.fr

References

[1] Dweik, R.A., *The great challenge for exhaled breath analysis: embracing complexity, delivering simplicity.* Journal of Breath Research, 2011. **5**(3): p. 030201.

[2] Horváth, I., et al., *A European Respiratory Society technical standard: exhaled biomarkers in lung disease.* European Respiratory Journal, 2017. **49**(4): p. 1600965.

[3] Steeghs, M.M.L., S.M. Cristescu, and F.J.M. Harren, *The suitability of Tedlar bags for breath sampling in medical diagnostic research.* Physiological Measurement, 2006. **28**(1): p. 73-84.

[4] Riess, U., et al., *Experimental setup and analytical methods for the non-invasive determination of volatile organic compounds, formaldehyde and NOx in exhaled human breath.* Analytica Chimica Acta, 2010. **669**(1): p. 53-62.

[5] Pleil, J.D. and A.B. Lindstrom, *Measurement of volatile organic compounds in exhaled breath as collected in evacuated electropolished canisters.* J Chromatogr B Biomed Appl, 1995. **665**(2): p. 271-279.

[6] Yang, H.-Y., et al., *Development of breath test for pneumoconiosis: a case-control study.* Respiratory research, 2017. **18**(1): p. 178-178.

[7] Scott-Thomas, A., M. Epton, and S. Chambers, *Validating a breath collection and analysis system for the new tuberculosis breath test.* Journal of Breath Research, 2013. 7(3): p. 037108.

[8] King, J., et al., *Dynamic profiles of volatile organic compounds in exhaled breath as determined by a coupled PTR-MS/GC-MS study.* Physiological measurement, 2010. **31**(9): p. 1169-1184.

[9] Kang, S. and C.L. Paul Thomas *How long may a breath sample be stored for at -80 °C? A study of the stability of volatile organic compounds trapped onto a mixed Tenax:Carbograph trap adsorbent bed from exhaled breath.* Journal of breath research, 2016. **10**, 026011

[10] Prado, C., P. Marín, and J.F. Periago, *Application of solid-phase microextraction and gas chromatography–mass spectrometry to the determination of volatile organic compounds in end-exhaled breath samples.* Journal of Chromatography A, 2003. **1011**(1): p. 125-134.

[11] Schallschmidt, K., et al., *Comparison of volatile organic compounds from lung cancer patients and healthy controls—challenges and limitations of an observational study.* Journal of Breath Research, 2016. **10**(4): p. 046007.

[12] Lawal, O., et al., *Exhaled breath analysis: a review of 'breath-taking' methods for off-line analysis.* Metabolomics, 2017. **13**(10): p. 110.

[13] Beccaria, M., et al., *Exhaled human breath analysis in active pulmonary tuberculosis diagnostics by comprehensive gas chromatography-mass spectrometry and chemometric techniques.* Journal of Breath Research, 2018. **13**(1): p. 016005.

[14] Beccaria, M., et al., *Preliminary investigation of human exhaled breath for tuberculosis diagnosis by multidimensional gas chromatography – Time of flight mass spectrometry and machine learning.* Journal of Chromatography B, 2018. **1074-1075**: p. 46-50.

[15] Available from: https://www.sigmaaldrich.com/content/dam/sigma-aldrich/docs/Supelco/General_Information/t402025.pdf.

[16] de Lacy Costello, B., et al., *A review of the volatiles from the healthy human body*. Journal of Breath Research, 2014. **8**(1): p. 014001.

[17] García, R.A., et al., *Volatile Organic Compounds Analysis in Breath Air in Healthy Volunteers and Patients Suffering Epidermoid Laryngeal Carcinomas*. Chromatographia, 2014. **77**(5): p. 501-509.

[18] Bean, H.D., C.A. Rees, and J.E. Hill, *Comparative analysis of the volatile metabolomes of Pseudomonas aeruginosa clinical isolates*. Journal of Breath Research, 2016. **10**(4): p. 047102.

[19] Shestivska, V., et al., *Quantitative analysis of volatile metabolites released in vitro by bacteria of the genus Stenotrophomonas for identification of breath biomarkers of respiratory infection in cystic fibrosis*. Journal of Breath Research, 2015. **9**(2): p. 027104.

[20] Shirey, R.E., *4 - SPME Commercial Devices and Fibre Coatings*, in *Handbook of Solid Phase Microextraction*, J. Pawliszyn, Editor. 2012, Elsevier: Oxford. p. 99-133.

[21] Yang, Q.J., et al., Comparing early liver graft function from heart beating and living-donors: A pilot study aiming to identify new biomarkers of liver injury. Biopharmaceutics & Drug Disposition. 2017. **38**(5): p. 326-339.

[22] Stagaard, R., et al., Abrogating fibrinolysis does not improve bleeding or rFVIIa/rFVIII treatment in a non-mucosal venous injury model in haemophilic rodents. 2018. **16**(7): p. 1369-1382.

[23] Roszkowska, A., et al., *Equilibrium ex vivo calibration of homogenized tissue for in vivo SPME quantitation of doxorubicin in lung tissue*. Talanta, 2018. **183**: p. 304-310.

[24] Jaroch, K., et al., *Untargeted screening of phase I metabolism of combretastatin A4 by multi-tool analysis*. Talanta, 2018. **182**: p. 22-31.

[25] Filipiak, W. and B. Bojko, *SPME in clinical, pharmaceutical, and biotechnological research – How far are we from daily practice?* TrAC Trends in Analytical Chemistry, 2019. **115**: p. 203-213.

[26] Zhang, S. and D. Raftery, *Headspace SPME-GC-MS Metabolomics Analysis of Urinary Volatile Organic Compounds (VOCs)*, in *Mass Spectrometry in Metabolomics*. 2014, Humana Press, New York, NY.

[27] Miekisch, W., et al., *Impact of sampling procedures on the results of breath analysis*. Journal of Breath Research, 2008. **2**(2): p. 026007.

[28] Castellanos, M., et al., *Breath gas concentrations mirror exposure to sevoflurane and isopropyl alcohol in hospital environments in non-occupational conditions*. Journal of Breath Research, 2016. **10**(1): p. 016001.

[29] Martin, A.N., et al., *Human breath analysis: methods for sample collection and reduction of localized background effects*. Analytical and Bioanalytical Chemistry, 2010. **396**(2): p. 739-750.

[30] Poli, D., et al., *Determination of aldehydes in exhaled breath of patients with lung cancer by means of on-fiber-derivatisation SPME–GC/MS*. Journal of Chromatography B, 2010. **878**(27): p. 2643-2651.

[31] Bruderer, T., et al., *On-Line Analysis of Exhaled Breath*. Chemical Reviews, 2019. **119**(19): p. 10803-10828.

[32] Kramer, J.K., et al., Conjugated linoleic acids and octadecenoic acids: Analysis by GC. 2001. **103**(9): p. 600-609.

[33] Marriott, P.J., et al., Multidimensional gas chromatography. 2012. **34**: p. 1-21.

[34] Nolvachai, Y., et al., *Multi-column trajectory to advanced methods in comprehensive two-dimensional gas chromatography.* 2018. **106**: p. 11-20.

[35] Sharif, K.M., et al., *The microfluidic Deans switch: 50 years of progress, innovation and application.* 2016. **82**: p. 35-54.

[36] Tranchida, P.Q., et al., *Modulators for comprehensive two-dimensional gas chromatography.* TrAC Trends in Analytical Chemistry, 2011. **30**(9): p. 1437-1461.

[37] Edwards, M., A. Mostafa, and T. Górecki, *Modulation in comprehensive two-dimensional gas chromatography: 20 years of innovation.* Analytical and Bioanalytical Chemistry, 2011. **401**(8): p. 2335-2349.

[38] Wilde, M.J., et al., *Breath analysis by two-dimensional gas chromatography with dual flame ionisation and mass spectrometric detection - Method optimisation and integration within a large-scale clinical study.* Journal of chromatography. A, 2019. **1594**: p. 160-172.

[39] Miekisch, W., et al., *Assessment of propofol concentrations in human breath and blood by means of HS-SPME-GC-MS.* Clin Chim Acta, 2008. **395** (1-2): p. 32-37.

[40] Ulanowska, A., et al., *Determination of volatile organic compounds in human breath for Helicobacter pylori detection by SPME-GC/MS.* Biomed Chromatogr, 2011. **25**(3): p. 391-397.

[41] Mellors, T.R., et al., *The volatile molecule signature of four mycobacteria species.* 2017. **11**(3): p. 031002.

[42] Mellors, T.R., et al., *A new method to evaluate macaque health using exhaled breath: A case study of M. tuberculosis in a BSL-3 setting.* 2017. **122**(3): p. 695-701.

[43] Buszewski, B., et al., Human exhaled air analytics: biomarkers of diseases. Biomedical Chromatography. 2007. **21**(6): p. 553-566.

[44] Almstetter, M.F., et al., Comprehensive two-dimensional gas chromatography in metabolomics. 2012. **402**(6): p. 1993-2013.

[45] Misra, B.B., et al., High-resolution gas chromatography/mass spectrometry metabolomics of non-human primate serum. 2018. **32**(17): p. 1497-1506.

[46] Yu, Z., et al., *Optimizing 2D gas chromatography mass spectrometry for robust tissue, serum and urine metabolite profiling.* Talanta, 2017. **165**: p. 685-691.

[47] Lamani, X., et al., *Determination of aromatic amines in human urine using comprehensive multi-dimensional gas chromatography mass spectrometry (GCxGC-qMS).* Analytical and Bioanalytical Chemistry, 2015. **407**(1): p. 241-252.

[48] Amann, A. and D. Smith, *Breath analysis for clinical diagnosis and therapeutic monitoring.* Thursday Dec 20, 2005: p. 201218.

[49] Cao, W. and Y. Duan, *Breath analysis: potential for clinical diagnosis and exposure assessment.* Clinical chemistry, 2006. **52**(5): p. 800-811.

[50] Cao, W. and Y. Duan, *Breath Analysis: Potential for Clinical Diagnosis and Exposure Assessment.* Clinical Chemistry, 2006. **52**(5): p. 800.

[51] Miekisch, W., J.K. Schubert, and G.F. Noeldge-Schomburg, *Diagnostic potential of breath analysis—focus on volatile organic compounds.* Clinica chimica acta, 2004. **347**(1-2): p. 25-39.

[52] Phillips, M., *Breath tests in medicine.* Scientific American, 1992. **267**(1): p. 74-79.

[53] Smith, D. and P. Španěl, *The challenge of breath analysis for clinical diagnosis and therapeutic monitoring.* Analyst, 2007. **132**(5): p. 390-396.

[54] Sethi, S., R. Nanda, and T.J.C.m.r. Chakraborty, *Clinical application of volatile organic compound analysis for detecting infectious diseases.* 2013. **26**(3): p. 462-475.

Chapter 5

Exploring the Mysteries of *Cannabis* through Gas Chromatography

María Teresa García-Valverde, Verónica Sánchez de Medina, Verónica Codesido, Jesús Hidalgo-García and Carlos Ferreiro-Vera

Abstract

In the last decades, cannabinoids, the active constituents of *Cannabis sativa* L., have been attracting a strong interest, regarding the health effects associated with the use of *Cannabis* and *Cannabis*-derived products. The progressive legalization of this species in several countries has prompted an increasing concern about the characterization and quantification of cannabinoids in diverse chemotypes of the plant, as well as the obtained final products. Therewith, Process and Product Quality Assurance (PPQA) becomes a mandatory practise to verify the Good Manufacturing Practices (GMP). Gas chromatography is one of the most used techniques in this sense due to its high attainable resolution. However, sample complexity and the thermal lability of cannabinoids hinder the analysis. In this chapter, a fully description of the recent advances in the *Cannabis sativa* L. analysis by gas chromatography will be presented, including different approaches that have come up to solve the obstacles encountered.

Keywords: *Cannabis sativa* L., hemp, complex matrix, cannabinoids, Terpenoids, gas chromatography, quality control assurance

1. Introduction

Cannabis sativa L. is the most thoroughly studied and widely used plant from Central Asia 3000 years before the Christian era. This annual dioecious plant has a complex chemical composition, including cannabinoids, terpenes, flavonoids, stilbenoids, fatty acids, alkaloids, carbohydrates, and polyphenols, among others. Therefore, this plant and its by-products have been widely used in different areas in the production of ropes, cloth, food or oil, being considered one of the most significant agricultural crops over the years. Additionally, since the chemical structure of Δ^9-tetrahydrocannabinol (Δ^9-THC), the main psychoactive compound in *Cannabis*, was identified by Mechoulam in the 60s, the studies about the active compounds of *Cannabis* increased dramatically and, owing to its bioactive components, *Cannabis* has been used for recreational, medicinal and scientific purposes [1]. In recent years, as a consequence of multiple scientific evidences, the number of countries around the world where the use of *Cannabis* is being legalized or decriminalized is continuously growing, like many Latin American countries (Uruguay, Peru, Venezuela, Chile, Colombia, Argentina, and Ecuador), different U.S. states, some countries of Europe (Italy, Czech Republic and Germany) and Canada. In this regulatory scenario, the most common practice is to use the Δ^9-THC content for the discrimination among industrial or medicinal varieties. However, the classification of *Cannabis* plants regarding this parameter varies depending on the legislation of the country of origin. While some countries like Switzerland, Uruguay or Colombia authorize 1.0% of Δ^9-THC for medicinal use, more restrictive legislation in the European Union limits this value to the 0.2% content. It should be noted that obviously only duly registered and certified varieties could be used [2]. When talking specifically about medical *Cannabis*, quality control of cultivars as well as standardization is a requirement for these applications as they are extremely important to ensure the health and safety of medical *Cannabis* users and patients, being necessary the development of reliable methods to quantify bioactive compounds from *Cannabis* [3, 4].

Therefore, several chromatographical techniques have been widely used for the identification and quantification of *Cannabis* constituents, being gas chromatography (GC), especially coupled to mass spectrometry (GC–MS), the most established technique in forensic and clinical toxicology analyses. Considering the similarities of their bioactive compounds in their physicochemical properties, chromatographic separation involves a mandatory step that may be usually considered as time consuming. For this reason, different strategies have been implemented increasing the resolution between isomers and overcoming this problem. In addition, the identification of these compounds is possible thanks to the advanced National Institute of Standards and Technology (NIST) mass-spectral database. In this manner, the analysis of *Cannabis* does not require highly sophisticated equipment being appropriate for laboratories with reduced instrumental availability. Nevertheless, different methods have been reported using high-resolution mass spectrometry allowing the untargeted analysis of *Cannabis* samples [5].

Generally, an internal standard (IS) is a nonendogenous compound, which is naturally similar to the target analyte. In this sense, this methodology is employed to increase the reproducibility of the quantification when there is a source of errors during sample analysis. These inaccuracies may be related to random and systematic errors, due to sample preparation or even the complexity of the sample, among other factors. The IS must be added to each sample in a constant amount, as well as to blank and calibration standards. In this manner, any deviation occasioned during the analysis of the sample will also affect the IS, correcting the result with the relation of both signals. For this reason, the selection of the IS is a critical step to guarantee the precision in the analysis. The stable-isotope labeled IS are the most suitable when a MS detector is used [6].

Furthermore, derivatization processes are generally required, improving the chromatographic resolution as well as peak shape of analytes [7]. For this reason, many derivatization agents have been applied, being the silylation approaches preferred. These derivatization agents are suitable to volatilize and improve mass fragmentation properties of active proton-containing groups. Commonly, N, O-bis (trimethylsilyl) trifluoroacetamide (BSTFA), N-methyl-N-(trimethylsilyl) trifluoroacetamide (MSTFA), trimethylchlorosilane (TMCS) or N-tert-butyldimethylsilyl-N- methyltrifluoroacetamide (MTBSTFA) are utilized [8]. Although this procedure turns out to be time-consuming, it is worthy since it provides increased sensitivity and reproducibility [6, 9]. Throughout this chapter, the authors will discuss the multiple analytical work modalities and methods that have been used to solve all the inconveniences found in the analysis of *Cannabis* samples using gas chromatography.

2. Cannabinoids

Cannabinoids are the major constituents of *Cannabis* plant, distributed on aerial surfaces/leaves and female inflorescences of the plants. These compounds are concentrated on resinous secretions produced by glandular trichomes [1, 10]. Their bioactive compounds are acquiring more notoriety because of their physiological effects as well as their medicinal properties, which are applied in the treatment of a wide range of diseases and disorders (e.g. multiple sclerosis, epilepsy, fibromyalgia). They are also employed for alleviating the pain induced by some treatment methods for various diseases, including cancer. Alongside being of great interest for the patient and medical communities, the European medicinal *Cannabis* market is expected to boom in the coming years. Consequently, the development of more efficient qualitative and quantitative methods for the analysis of these compounds is required [11].

Cannabinoids may be classified into three groups based on their source of production, *viz.*, endocannabinoids, phytocannabinoids and synthetic cannabinoids [12]. There is an extensive list of these compounds, hence only some of them are depicted in **Figure 1**.

2.1 Endocannabinoids

Endocannabinoids (ECs) are defined as endogenous lipids that are involved in many physiological and pathological conditions, regulating neurological disorders. They are characterized by their cannabimimetic features, activating the cannabinoid receptors (CB1 and CB2) as well as other receptors. For this reason, the development of reliable methodologies to determine ECs levels is mandatory in order to understand the role of these lipid metabolites. Different ECs have been widely analyzed by GC–MS and LC–MS, like arachidonoyl ethanolamide (anandamide, AEA) and 2-arachidonyl glycerol (2-AG) in human biosamples, which are the two most widely studied [13]. Additionally, other ECs like virodhamine and noladin ether has been detected. These compounds present physiological properties comparable to natural and synthetic exogenous cannabinoids. However, the main drawback found in their determination is their low concentration, acute instability of some endogenous compounds as well as the limited sample availability. For this reason, different extraction techniques like liquid–liquid extraction (LLE) or solid phase extraction (SPE) have been used for the quantification of ECs from different samples, to increase the analyte concentration. Moreover, microextraction

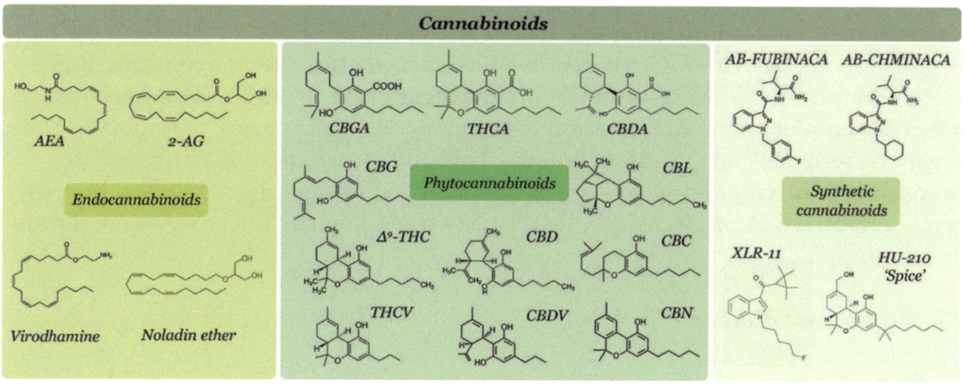

Figure 1.
Classification of cannabinoids based on their source of production.

approaches like solid phase microextraction (SPME) have been employed in the ECs analysis, reducing the required amount of sample [14].

Several strategies using GC–MS have been described in the literature for the determination of ECs from biological samples [15–17]. In this regard, different derivatization reagents like BSTFA or methyl ester of the methoxim dimethyiso-propylsilyl (DMiPSi) have been employed, in order to increase the stability of these compounds during their analysis. Additionally, different ionization modes, including electron ionization (EI), positive chemical ionization (PCI) and negative chemical ionization (NCI, also denoted as NICI, ECCI, ECNI and ECNICI) have been utilized. Such is the case of the method reported by Kayacelebi et al., based on the indirect evaluation of 2-AG by the analysis of arachidonic acid (AA) and prostaglandin E_2 (PGE_2), which are hydrolysis products of the EC. This approach employs GC–MS and GC–MS/MS using the NCI mode. This technique provides enhanced selectivity in comparison to EI since only analytes with high electron capture capacity or high electron affinity may be ionized, thereby eliminating potential sample matrix interferences and allowing detections at very low concentration levels (ng/L). This consideration provides an advantage in contrast to LC–MS, since sample matrix is usually one of the major issues in this methodology. In NCI, low energy electrons (around 2 eV) are produced by collision of the reagent gas (generally methane, ammonia or carbon dioxide) with electrons emitted from the filament. The resulting electrons are captured by the analyte producing stable molecular anions. Although only specified analytes fulfill the characteristics to be used by this technique, derivatization processes provide wider applicability to this unusual ionization mode. For instance, the perfluoro group presents high electron affinity, being suitable for NCI. In this manner, an esterification process of AA and PGE_2 was accomplished using pentafluorobenzyl bromide (PFB-Br) as derivatization agent, allowing the controlled fragmentation of both metabolites and accordingly, resulting in a highly sensitive quantitative method [18]. Additionally, PCI has been used for the determination of AEA by GC–MS using TMS as derivatization agent. As with NCI, PCI is based on the generation of protonated molecules by collision of the reagent gas with an electron emitted from the filament. The ionized reagent gas reacts with the analyte, resulting in a positive adduct ion. In both PCI and NCI, information about the molecular weight of the analyte is provide, unlike EI where higher energy electrons collide with analytes creating fragmented positive ions and other species. Even though this methodology may appear simpler, the selectivity reached by NCI outweighs the benefits of PCI, as well as the weaknesses of the technique itself, *viz.*, dissociative reactions by high energy electrons, analyte response suppression by oxygen or water contamination and affected reaction yield and fragmentation behavior due to high ionization chamber temperature [19]. These strategies are summarized in **Table 1**.

Despite the fact that GC–MS has been extensively utilized to determine ECs metabolites from biological samples, in the recent years, LC–MS/MS has become the generally employed instrumentation in the analysis of ECs, being considered in this sense the reference analytical technique. This preference may be attributed to increased sensitivity and selectivity as well as high-throughput capability. In addition, derivatization processes of ECs may be avoided, which are laborious and time-consuming [20].

2.2 Phytocannabinoids

Phytocannabinoids are naturally occurring cannabinoids found in the *Cannabis* plant. These compounds present a similarity in their chemical structure since they are constituted by a C_{21} structural feature with alterations in the length of their side

CANNABINOIDS	Sample treatment					Instrumental technique				Analytical features					
	Sample	Amount of sample	Metabolite	Derivatization	Extraction	Technique	Ionization source	Column	Analysis Time (min)	LOD	LOQ	Precision (RSD, %)		Recovery (%)	Ref.
												Intra-assay	Inter-assay		
ENDOCANNABINOIDS	Plasma	1.5 mL	AEA and 2-AG	BSTFA/TMCS	LLE	GC–MS	PCI	DB-1 MS	16.2	0.3–0.5 ng/mL	0.35–1 ng/mL	<11	10.6–14.1	42.7–72.2	[13]
	Blood	0.2–5 mL	AEA and 2-AG	DMiPSi	SPE	GC–MS	EI	DB-1	<20	—	—	—	—	—	[15]
	Mouse brain	—	AEA and 2-AG	BSTFA	SPE	GC–MS	PCI	Rtx-5MS	11	0.01 pmol	0.2 pmol	2.5		70.9	[16]
	Mouse brain	30 mg	AEA, 2-AG and PEA	BSTFA	SPE	GC–MS	CI	CP-Sil8 CB	<30	—	—	—	—	—	[17]
	Dog liver	—	2-AG (AA and PGE2)	PFB-Br	—	GC–MS and GC–MS/MS	NCI	OPTIMA-17MS	32	—	10 nM / ≤0.1 μM	2.3–12.9		—	[18]
	Tissue	50–100 mg	NAE (PEA, SEA, AEA)	BSTFA	SPE	GC–MS and GC–MS/MS	EI	Rtx-5MS	—	—	—	—	—	—	[19]
PHYTOCANNABINOIDS			11-OH-THC							0.03 pg./mg	0.1			88	[27]
	Hair			MSTFA	LLE	GC–MS/MS	EI	DB5-MS				9–14	5–9		[27]
			THC, CBD and CBN							0.3–1.4	0.9–4.7 pg./mg			68–97	[27]
	Plant (buds, leaves and stems)	0.2 g	CBD, THC and CBN	—	HCEE	GC-,S	EI	Zebron ZB-5MS	30–60	0.05–0.09 mg/L	0.17–0.30 mg/L	5.6–7.4		97–99	[28]

CANNABINOIDS	Sample treatment					Instrumental technique				Analytical features					
	Sample	Amount of sample	Metabolite	Derivatization	Extraction	Technique	Ionization source	Column	Analysis Time (min)	LOD	LOQ	Precision (RSD, %)		Recovery (%)	Ref.
												Intra-assay	Inter-assay		
	Cannabis seed	5 g	THC, CBD, CBC, CBG and CBN	—	PHWE	GCXGC-QTOF-MS/MS	EI	BPX5	30	—	—	2.87–4.5	1.32–5.60	89–45–92-56	[29]
	Hair	50 mg	THC, CBD, CBN and 11-OH-THC	MSTFA	LLE	GC–MS/MS	EI	DB5-MS	14	0.303–1.4 pg./mg	0.1–4.7 pg./mg	1.84–8.8	2.78–14.05	68–97	[31]
			THC-COOH	PEPA:HFIP			NCI		8						
	Serum and plasma	1 mL	THC, THC-COOH and 11-OH-THC	MSTFA	SPE (AXS)	GC–MS	EI	OPTIMA® 5 MX Accent	<30	0.15–2 ng/mL	0.3–3.3 ng/mL	0.8–4.8	1.9–6.1	76.9–96.5	[32]
	Human breast milk	0.05 mL	THC, CBN and CBD	—	SPME	HS-SPME-GC–MS	EI	HP-5MS	16	10 ng/mL	20 ng/mL	6.5–13.3	2.1–9.2	90.9–118	[33]
	Hair	25 mg	THC-COOH, OH-THC, THC, CBD and CBN	MSTFA	SPE	GC–MS/MS	EI	Zebron ZB-5MSi	68	0.2–2 pg./mg	0.5–5 pg./mg	1.6–5.5	2.3–6.6	19–79	[34]
	Human serum	1 mL	THC, THC-OH, THCA, CBD, CBDA and CBG	MSTFA	LLE	GC–MS	AP	HP-5MS	12	0.05–0.9 ng/mL	0.2–3 ng/mL	0.2–14.6	6.9–14.9	>82	[35]
	Plant (roots, stems, buds and leaves)	100 mg	THCV, CBD, CBC, Δ8-THC, Δ9-THC, CBG, CBN, CBDA, THCAA and CBGA	BSTFA	—	GC-FID	—	DB-1MS	17.5	0.11–0.19 µg/mL	0.34–0.56 µg/mL	0.19–16.79		—	[36]

CANNABINOIDS	Sample treatment					Instrumental technique				Analytical features					
	Sample	Amount of sample	Metabolite	Derivatization	Extraction	Technique	Ionization source	Column	Analysis Time (min)	LOD	LOQ	Precision (RSD, %)		Recovery (%)	Ref.
												Intra-assay	Inter-assay		
	Buccal swabs	5 mg	THC, Δ8-THC, CBN, CBD, CBC, CBG, and CBDV	MSTFA	SPME	HS-SPME-GC–MS	EI	Rxi-35 Sil	12	< 0.04 μg/mg	—	—	—	—	[37]
	Plant (flowers)	—	CBD, THC and CBN	—	SBSE	GCXGC-QTOF-MS	EI	Rxi-5MS	93	0.02–0.15 μg/mL	0.05–0.51 μg/mL	8.8–19.3		—	[5]
	Standard mixtures	—	THCV, CBD, CBC, Δ8-THC, Δ9-THCA, 11-hydroxy-Δ9-THC and 11-nor-9-carboxy-Δ9-THC	BSTFA+1% TMCS	—	GC-UV/VUV	—	Rtx-5	15	5–10 mg/L	—	—	—	—	[11]
	Standard mixtures	—	62 SCs	—	—	GC–MS	PI	DB-5MS	—	—	—	—	—	—	[41]
SYNTHETIC CANNABINOIDS	Seized plant	150 mg	15 SCs	—	—	GC–MS	cokl-EI	DB-5MS	>15	—	8–133 μg/L	2.9–7.3		—	[42]
	Seized plant	—	34 SCs	—	—	GC–MS/MS	EI	DB-5MS	>30	19.9–68.8 ng/mL	—	—	—	—	[43]
							CI			91.1–162.9 ng/mL					

CANNABINOIDS	Sample treatment					Instrumental technique				Analytical features					
	Sample	Amount of sample	Metabolite	Derivatization	Extraction	Technique	Ionization source	Column	Analysis Time (min)	LOD	LOQ	Precision (RSD, %)		Recovery (%)	Ref.
												Intra-assay	Inter-assay		
	Standard mixtures		6 SCs	—	—	GC–MS	EI	Rtx-200	21	—	—	—	—	—	[44]
						GC-IR	—	BPX5							
	Plant	10 mL	11 SCs	—	—	GC–MS	EI PCI NCI	HP-5MS	30	—	—	—	—	—	[45]
	Urine							J&W scientific	21	1–5 μg/L	5 μg/L	6.1–16.2		54–98.2	[6]
		2 mL	29 SCs	BSTFA	UADLLME	GC–MS	EI	5% phenyl-methylsilicone		1–5 μg/L	5 μg/L	0.5–19.7		93.8–105.3	[6]
	Blood														
	Plant	20 mg	44 SCs	—	SPME	HS-SPME-	EI	DB-5 ms	23	—	—	—	—	—	[46]
	Blood	500 μL				GC–MS									
	Saliva	455 μL	5 SCs	—	MEPS	GC–MS	EI	HP-5 MS	26	10–20 μg/L	30–60 μg/L	3.6–8.9		62–124	[47]

2-AG: 2-arachidonyl glycerol; AA: arachidonic acid; AEA: anandamide; AXS: anion exchange sorbent; BSTFA: N,O-bis (trimethylsilyl) trifluoroacetamide; CBC: cannabichromene; CBD: cannabidiol; CBDA: cannabidiolic acid; CBDV: cannabidivarin; CBG: cannabigerol; CBGA: cannabigerolic acid; CBN: cannabinol; DMiPSi: methyl ester of the methoxim dimethyisopropylsilyl; EI: electron ionization; FID: flame ionization detector; GC: gas chromatography; HCEE: hard cap espresso extraction; HFIP: hexafluoroisopropanol; IR: infrared; LLE: liquid–liquid extraction; MEPS: micro extraction by packed sorbent; MS: mass spectrometry; MSTFA: N-Methyl-N-(trimethylsilyl) trifluoroacetamide; NAE: N-acylethanolamines; NCI: negative chemical ionization; PCI: positive chemical ionization; PEA: palmitoylethanolamide; PFB-Br: pentafluorobenzyl bromide; PFPA: pentafluoropropionic anhydride; PGE_2: prostaglandin E_2; PHWE: pressurized hot water extraction; PI: photoionization; SBSE: stir bar sorptive extraction; SEA: stearoylethanolamide; SPE: solid phase extraction; SPME: solid phase microextraction; THC: tetrahydrocannabinol; THCAA: tetrahydrocannabinolic acid A; TMCS: trimethylchlorosilane; QTOF: quadrupole time-of-flight; UADLLME: ultrasound-assisted dispersive liquid–liquid microextraction; UV/VUV: ultraviolet/vacuum ultraviolet.

Table 1.
Analytical platforms for the analysis of cannabinoids in Cannabis *samples.*

chain (C1-C5) attached to the aromatic ring. Cannabigerolic acid (CBGA) is a precursor molecule, which can be converted through a series of enzymatic reactions into Δ^9-tetrahydrocannabinolic acid (Δ^9-THCA), cannabidiolic acid (CBDA), or cannabichromenic acid (CBCA). Acidic phytocannabinoids may be subsequently decarboxylated into their corresponding active neutral form by decarboxylation processes, losing a COOH group. This process may occur over time, under heating or alkaline conditions [10]. Consequently, the previously mentioned cannabinoids will be transformed into cannabigerol (CBG), Δ^9-THC, cannabidiol (CBD) and cannabichromene (CBC). All these cannabinoids are characterized by a composition of five carbon atoms in their side chain. Additional compounds have been detected varying the length of their side chain, thereby varinoids compounds, such as tetrahydrocannabivarinic acid (THCVA), cannabidivarinic acid (CBDVA), cannabidivarin (CBDV) and tetrahydrocannabivarin (THCV) are naturally produced in the plant, with a three-term alkyl chain. Some other compounds, like Δ^8-tetrahydrocannabinol (Δ^8-THC), cannabinol (CBN), cannabicyclol (CBL), cannabinolic acid (CBNA), and cannabicyclolic acid (CBLA) have also been reported. Additionally, the exposition to environmental agents like oxygen may lead to the alteration of these compounds. The oxidative degradation of Δ^9-THC to CBN under oxygen exposition is a representative example of this condition [21].

Currently, almost 150 phytocannabinoids have been detected in *Cannabis* plant, although most of them have never been isolated or characterized. Therefore, the list is still under construction since new cannabinoids are continually being discovered [22]. Recently, new phytocannabinoids have been detected, *viz.*, cannabidibutol (CBDB) and Δ^9-tetrahydrocannabutol (Δ^9-THCB), characterized by four carbon atoms in their side chain, as well as cannabidiphorol (CBDP) and tetrahydrocannabiphorol (Δ^9-THCP), being the seven-term homologs of CBD and Δ^9-THC, respectively [23]. Δ^9-THCP proved to be as active as Δ^9-THC but at lower doses, being necessary further analyses to evaluate the pharmacological effect of this potent phytocannabinoid.

For many years, *Cannabis* has had a negative connotation due to the psychotropic effects associated with Δ^9-THC. However, researchers have been working for a long time against this situation due to the beneficial health components found in the plant. All the bioactive components contribute in different manner to health, being used as treatment against different diseases. Among them, CBD has been the most widely studied phytocannabinoid, since it presents a series of medical properties like antioxidant, anti-inflammatory, antibacterial, anti-proliferative and neuroprotective effects. However, all this flood of information has diverted the attention from some other minor cannabinoids with more remarkable and interesting characteristics. In this sense, CBDV, CBG and Δ^9-THCV have been barely investigated since the low concentration of these cannabinoids in the plant hinder their isolation for further studies. Nevertheless, there is evidence of the anti-inflammatory and anti-proliferative properties of CBG and CBC, together with a significant antibacterial activity [24, 25]. Thus, the genetic selection of *Cannabis* varieties rich in other minor phytocannabinoids would confer the opportunity of extract these compounds.

From a medical point of view, the administration of these compounds has been normally accomplished by *Cannabis* oil extracts, which are produced from dried *Cannabis sativa* L. inflorescences incorporated in common edible oils (*e.g.*, olive or sunflower), sometimes using these oils as extraction media [26]. Thus, *Cannabis* extraction has been thoroughly studied to obtain highly concentrated content of cannabinoids and other beneficial components. Different organic solvents like ethanol, methanol, acetone, or hexane have been utilized for this purpose, although more natural approaches, like water, have been also evaluated. An exhaustive control of the potency as well as the standardization of different plant batches may be accomplished to guarantee the correct dosage of phytocannabinoids in the final oil. Accordingly, analyses of the extracts resulted of the maceration in olive oil have been performed by GC coupled to flame ionization detector (FID) and GC–MS, as a way of comparison (see **Table 1**). Results were comparable in terms of precision, accuracy and linearity [27]. Recently, an innovative and rapid method has been reported to the extraction of cannabinoids, using the hard cap espresso extraction (HCEE) methodology coupled to GC–MS [28]. This approach allows the quantitative extraction of THC, CBD and CBN from seized *Cannabis* samples in only 40 s, using 100 mL of isopropanol and GC–MS determination. The obtained results were compared with those acquired by the ultrasound-assisted extraction (UAE) reference method, being both approaches statistically comparable. Similarly, a green extraction method, pressurized hot water extraction (PHWE), has been used to extract cannabinoid compounds from *Cannabis sativa* L. seeds. This technique, based on the supercritical fluid extraction, is an alternative to CO_2 that enables the extraction of polar and semi-polar bioactive compounds from *Cannabis* seeds. In this case, GCxGC-TOF-MS methodology allowed the identification of cannabinoids without the need of using standards [29]. Considering that these concentrated cannabinoids extracts will be used in the pharmaceutical industry, an exhaustive control is mandatory to ensure consumer safety according to the GMP system [30].

These compounds are metabolized in the organism after consumption, being possible the detection of their metabolites. Analysis of these chemical by-products is an excellent solution to distinguish between passive drug exposure and active consumption. In this sense, Δ^9-THC is rapidly adsorbed and metabolized to 11-nor-Δ^9-tetrahydrocannabinol-9-carboxylic acid (Δ^9-THC-COOH) and 11-hydroxy-Δ^9-THC (11-THC-OH), using these compounds as *Cannabis* biomarkers of the psychoactive form. However, the low concentration of these metabolites in biological samples implies the use of high sensitive equipment like GC–MS/MS or LC–MS/MS as well as the extraction and preconcentration of the analytes before the analysis, which is time and solvent consuming, especially when liquid–liquid extraction (LLE) is utilized [31]. Alternatives approaches to LLE has been reported, including anion exchange sorbent (AXS) and SPME in combination with head space GC–MS (HS-SPME-GC–MS) [32, 33].

Recently, Beike et al. reported an automated process of sample preparation, which includes extraction, clean up and derivatization of cannabinoids, prior to GC-EI-MS/MS. The determination of THC as well as its metabolites Δ^9-THC-COOH and 11-THC-OH in hair samples was accomplished using a SPE cartridge attached to the module of the MultiPurposeSampler (MPS) autosampler from GERSTEL, allowing the complete automation of the extraction process with relative standard deviation better than 7% [34]. In the same way, distinct GC ionization sources have been applied to achieve enhanced sensitivity, like atmospheric pressure (AP) source. This recently commercialized modality provides better ionization results than electron ionization (EI) or chemical ionization (CI) sources, enhancing selectivity and, consequently, sensitivity. AP source has been used for the quantitative determination of cannabinoids in serum samples at trace levels [35]. This softer ionization technique is based on the ionization of compounds by corona discharge in the presence of nitrogen, operating at atmospheric pressure. However, EI has been employed as the common ionization source for the analysis of *Cannabis*. This modality is characterized by its higher ionization energy, which produces extensive fragmentation of the molecular ion, thus reducing the possibility of using the better known and higher mass precursor ions (typically used in ESI-LC–MS) in selected reaction monitoring mode (SRM) and, on the other hand, the presence of higher mass precursor ions in CI improves the potential for lower detection limits in CI SRM as compared to EI SRM.

Phytocannabinoids have been widely analyzed by GC techniques, using predominantly FID and MS as detectors. One of the main shortcomings of the analysis of these compounds by GC is the direct analysis of acidic phytocannabinoids, which undergoes decarboxylation in the injector port due to the high temperatures. This drawback can be avoided by derivatization processes, which increase the chemical stability of the acidic compounds [9, 36]. For this reason, generally liquid chromatography (LC) is preferred since derivatization step is avoided. Nevertheless, greater sensitivity is reached with GC methods, still being the instrumental technique of choice. Franklin et al. reported the headspace derivatization of cannabinoids during HS-SPME step. To this end, 5 μL of derivatization reagent, MSTFA, was inserted in a 20 mL sample vial with the aid of a vial insert. The presence of MSTFA effectively derivatized the neutral cannabinoids enhancing the sensitivity of the determination. However, an earlier derivatization of the acidic cannabinoids is necessary since the performance of the headspace derivatization in this case was not so effective [37].

An additional problem associated with the determination of cannabinoids by GC–MS is the difficulty to distinguish between isomers without compromising analysis time. This is the case of Δ^9-THC, which presents four isomers: Δ^8 -THC, CBD, CBC, and CBL. Some strategies have been developed to overcome this problem, like the inclusion of two-dimensional GC (GCxGC) technique, which allows the identification of complex samples, identifying biomarkers and cannabinoid isomers [5]. In this case, stir bar sorptive extraction (SBSE) coated with poly-(dimethylsiloxane) (PDMS) was utilized allowing the preconcentration of a wide range of compounds with different volatilities. The retained metabolites were directly desorbed in a GCxGC–MS system, resulting in a green analytical treatment procedure. Although isomers resolution is solved with this system, analysis time remains a problem since it is necessary to perform an extraction step for at least 60 min in addition to the 93 min per run. Nevertheless, the automatization of the process may reduce human involvement and therefore systematic and random errors could be limited. Conversely, the use of GC with vacuum UV spectroscopy (VUV) could be a potential solution since differentiation among isomers may be solved in shorter analysis times (<15 min). This technology analyses compounds in the UV/VUV spectral range (120–240 nm) and is based on the excitation of chemical bonds [11]. This strategy has been utilized for the determination of different cannabinoids and their metabolites. The deconvolution of co-elution peaks allows the reduction of the chromatographic time and therefore the analysis process. However, this solution constitutes an improvement at the expense of sensitivity.

2.3 Synthetic cannabinoids

Synthetic cannabinoids (SCs) are a class of designed drugs that simulates the effects of Δ^9-THC towards cannabinoid receptors. Originally, the synthesis of these compounds was intended with therapeutic purposes as a treatment of pain, being later introduced in the recreational drug market [38]. Some of the best-known SCs include AB-FUBINACA (N-(1-amino-3-methyl-1-oxobutan-2-yl)-1-(4-fluorobenzyl)-H-indazole-3-carboxamide), AB-CHMINACA (N-(1-amino-3-methyl-1-oxobutan-2-yl)-1-(4-fluoro-benzyl)-1H-indazole-3-carboxamide), XLR-11 ((1-(5-fluoropentyl)-1H-indol-3-yl)(2,2,3,3-tetramethylcyclopropyl) methanone) and HU-210 (1,1-Dimethylheptyl-11-hydroxy-tetrahydrocannabinol), commercialized in many European and US countries as 'Spice', among others. Some of these cannabinoids are represented in **Figure 1**. There is a real risk consuming this type of drugs since their chronic abuse often leads to dangerous side effects as well as toxicity and, even in some cases, resulting in death. In addition, new psychoactive substances are constantly emerging, presenting unknown and unexpected effects. Consequently, further research is needed in this field entailing an enormous challenge, since it requires a continual actualization of the analytical techniques with the lack of available chemical reference standards. Chromatographical techniques like GC has become an important tool in this field to determine the prevalence and to assess the risks of SCs [39]. An additional problem is related to the chemical structure of these compounds, which is not well defined and therefore hinder even further their detection. While some of them, denoted as classical SCs, are structurally like natural cannabinoids, others named nonclassical and hybrid SCs present different structural features. For this reason, researchers studied fragmentation pathways of these compounds to facilitate their identification by GC–MS [40]. Although this process appears simple, some compounds result in similar mass spectra, impeding the identification of the SCs. Tandem and high-resolution MS may solve this obstacle, but generally, these equipments are not easily accessible. Therefore, more economically feasible solutions are needed. One cost-effective solution to this issue could be the utilization of different ionization modes, such as photoionization (PI). Akatsu et al. reported the utilization of GC-PI-MS for the identification of 62 SCs. This technique allows the detection of stable neutral compounds with low ionization threshold using a radical cation produced by ultraviolet light radiation which traps an electron from the target molecule. This effect is not possible with EI, which is an adequate technique for relatively high ionization threshold. In this way, with the combination of both ionization modes, it would be possible the identification of SCs [41]. Additionally, alternative ionization sources as cold-EI or CI have been described to overcome the SCs lability towards EI. These techniques increase the abundance of molecular ions allowing the discrimination between structurally-related compounds enhancing at the same time the limit of detection (LOD) and quantification (LOQ) of the procedure (see **Table 1**) [42, 43]. Alternatively, the combination of the GC-EI-MS and GC-IR techniques allowed the separation of six SCs regioisomers with excellent resolution results [44]. Additionally, Umebachi et al. reported the combination of GC–MS using EI, PCI and NCI for the structural elucidation and identification of different SCs in herbal products. The fragmentation of the indole and indazole SC structures was monitored with the different ionization modes, using methane as reagent gas. Although EI gives enough information about the fragmentation pattern of the molecules, PCI and NCI provide the $[M+H]^+$ and $[M-H]^-$ fragment ions, which could be used for the molecular weight estimation [45].

On the other hand, complexity of samples may cause a reduction in the sensitivity of the determination as well as to affect the chromatographic system, particularly the stationary phase of the analytical column. Consequently, routine methods for identification of new psychoactive substances generally require quick, cheap and simple extraction methods. Scientists have been working for many years addressing this deficiency, providing new and upgraded extraction techniques to remove the undesired sample interferences. Mercieca et al. have developed a simple method to determine SCs and their metabolites in urine and blood samples. The procedure included ultrasound-assisted liquid–liquid microextraction (UALLME) and silylation derivatization coupled with GC–MS, being able to effectively identify 36 SCs and related metabolites with acceptable accuracy and precision results [7]. In the same way, HS-SPME-GC/MS has been utilized to determine 40 SCs in herbal mixtures as well as blood samples. The utilization of the SPME methodology allows the reduction of blood sample volume to 500 μL, of great advantage in the forensic analysis [46]. Recently, a new configuration using microextraction by packed sorbents (MEPS) to isolate five SCs from oral biofluids has been reported. This method allows the quantification of these compounds in the low μg/L level by GC-EI-MS, requiring just 500 μL of sample. MEPs configuration allows the potential automatization of the process for routine analysis, improving the method precision [47].

3. Terpenes

Apart from cannabinoids, *Cannabis sativa* L. is rich in terpene compounds, being the major components of the *Cannabis* essential oils (EO). In the same way to cannabinoids, hemp EO is secreted by glandular trichomes, which are presented in the leaves and inflorescences of the plant. Terpenes represent the volatile fraction, responsible for the characteristic smell of the plant [48]. Moreover, their health benefits and antimicrobial and insecticidal properties have recently increased the interest in the hemp EO. In addition, some studies reveal a synergistic action of terpenes with cannabinoids, known as 'entourage effect', in the treatment of pain, inflammation, depression, anxiety, addiction, epilepsy, cancer, and infections [4]. For this reason, more than 120 terpenes have been identified in *Cannabis* using GC-FID and GC–MS. Furthermore, recent studies have revealed that the terpenoid profile is a useful tool since compositional differences may be used to distinguish between *Cannabis* cultivars that have similar cannabinoid content [49, 50]. Terpenes may be classified according to their number of carbon atoms and the isoprene residues present in their structure as monoterpenes, diterpenes, triterpenes and sesquiterpenes, being the monoterpenes α-pinene, β-myrcene and α-terpinolene, and the sesquiterpenes β-caryophyllene and α-humulene the most abundant compounds in *C. sativa* L (**Figure 2**).

Some strategies have been developed to elucidate the terpene profile in *Cannabis* samples (see **Table 2**). This is the case of HS, coupled in some cases to SPME [51, 52], and two-dimensional GCxGC [5]. Meiri et al. reported the simultaneous analysis and quantification of 93 terpenoids in *Cannabis* inflorescences by static headspace (SHS) coupled to GC–MS/MS at very low μg/L level [53]. SHS allows the direct analysis of plant materials without sample treatment, which may affect negatively to the terpenoid content. Additionally, direct injection of these complex sample matrices is not recommended because coextracted interferences potentially hinder terpenes determination. On the other hand, low-volatility compounds may

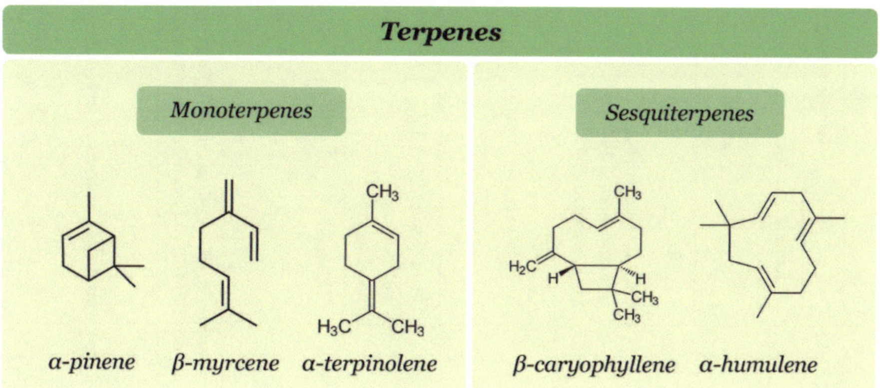

Figure 2.
The most abundant terpenes in Cannabis sativa *L.*

TERPENES	Sample treatment			Instrumental technique				Analytical features					Ref.
Sample	Amount of sample	Derivatization	Extraction	Technique	Ionization source	Column	Analysis time (min)	LOD	LOQ	Precision (RSD, %)			
										Intra-assay	Inter-assay	Recovery (%)	
EO	—	—	SPME	HS-SPME-GC–MS	EI	DB-5	60	—	—	—	—		[51]
Inflorescences and macerated oils	100 mg	—	SPME	HS-SPME-GC–MS	EI	Rtx-Wax	33	—	—	—	—		[52]
Inflorescences	5 mg	—	—	SHS-GC–MS/MS	EI	DB-35MS UI	74	0.001–0.123 μg/mL	0.002–0.374 μg/mL	<11.4	<21.9	81.2–119.6	[53]
Plant (leaves, inflorescence and seeds)	1 g	MSTFA:TMCS, 99:1, v/v	—	fast-GC-FID	EI	Restek RTX-5	<16	0.03–0.27 μg/mL	0.10–0.89 μg/Ml	3.59	7.82	77.52–107.10	[2]
Cannabis oil	400	—	SPME	HS-SPME-GC–MS	EI	HP-5-cross-linked poly-5% diphenyl-95% dimethyl polysiloxane	63	—	—	—	—		[55]

***EI:** electron ionization; **EO:** essential oil; **FID:** flame ionization detector; **GC:** gas chromatography; **HS:** head space; **MS:** mass spectrometry; **MSTFA:** N-Methyl-N-(trimethylsilyl)trifluoroacetamide; **SHS:** static head space; **SPME:** solid phase microextraction; **TMCS:** trimethylchlorosilane.*

Table 2.
Analytical platforms for the analysis of terpenes in Cannabis *samples.*

be retained in the injection port requiring more frequent maintenance [54]. Nevertheless, some strategies have been reported by direct injection using fast-GC-FID, analyzing 29 terpenes and CBD from leaf samples in less than 16 min [2]. Despite higher amount of sample is used in the later approach, the sensitivity is affected, being approximately 10-fold lower than the GC–MS/MS procedure, probably due to the utilization of different detection systems. On the other hand, the precision is improved after the derivatization with MSTFA:TMCS.

Sample storage plays an important role in the terpene analysis, having monoterpenes a tendency to evaporate more easily than sesquiterpenes. For this reason, special attention is required when analyzing these compounds. Some strategies have been evaluated such as using frozen samples instead of dried plant material or dynamic maceration at room temperature of plant inflorescences, providing excellent results [9, 55]. Additionally, oxygenated terpenoids, because of the presence of light or oxygen, may reveal some problems during plant storage and processing. Thus, the terpene profile provides interesting information about the EO or the starting plant material, such as the aging of the product, bad storage conditions or quality and breeding conditions. Terpene analysis provides these kind of information to take into consideration [25].

4. Other compounds

As it was described before, *Cannabis sativa* L. is mainly composed by cannabinoids and terpenes. However, some studies were focused on the determination of other minor compounds, like fatty acids and carbohydrates (see **Table 3**). Devi et al. reported the analysis of hemp oil by GC–MS for the determination of fatty acids and other compounds to be exploitable as biofuel [56]. In their study, different solvent extraction techniques were evaluated, such as supercritical fluid extraction (SFE), Soxhlet (SOX) and percolation (PER), among others. Although SOX is the best energy-efficient process, SFE is preferable in terms of purity of the obtained oil. However, alternative strategies might be implemented since large amount of solvents and extended extraction times make this approach economically and environmentally unattractive. On the other hand, an innovative method using hemp steam treatment has been described [57]. Steam and hot water were used to extract different compounds from diverse parts of hemp plants, such as stalk and leaves, using LLE and GC–MS for this purpose. The procedure allows the identification of several compounds like carbohydrates, fatty acids and aldehydes, among others. This procedure represents an improvement in comparison to the Devi et al. approach since analysis time is greatly reduced as well as solvent consumption. Additionally, Delgado-Povedano et al. reported two analytical platforms (GC-TOF/MS) and (LC-QTOF-MS/MS) for the untargeted characterization of *Cannabis* extracts, identifying a wide range of families including lipids, flavonoids, and amino and organic acids, among others, in multiples varieties of medicinal *Cannabis* [50]. The GC platform allowed the identification of 134 compounds in plant extracts in comparison with the 46 compounds identified by LC.

In the countries where *Cannabis* has been legalized for medical purposes, the governing agencies have established a series of strict regulations to guarantee the quality, safety, and usefulness of *Cannabis* products. Generally, these guidelines include the necessary quality control of pesticides, residual solvents, mycotoxins, heavy metals, and microbes to agree with the GMP system. The first two contaminants are widely controlled in the pharmaceutical industry using GC methodologies for their analysis. Short- and long-term adverse health effects are associated with the exposure to these contaminants through consumption of *Cannabis* products [58].

OTHER COMPOUNDS	Sample treatment					Instrumental technique				Ref.
	Sample	Amount of sample	Target	Derivatization	Extraction	Technique	Ionization source	Column	Analysis time (min)	
	Seed oil	—	Fatty acids	—	SFE/SOX/PER/ULT/UTS/STU	GC–MS	EI	DB-5MS	60	[56]
	Plant	2 mg	Carbohydrates	TMCS:HMDS	LLE	GC–MS	EI	HP-5 ms	28	[57]
	Plant (leaves and inflorescences)	100 mg	Untargeted	BSTFA + TMCS	—	GC-TOF	EI	DB-5MS-UI	53	[50]
	—	—	Pesticides	—	—	GC–MS/GC–MS/MS	EI	ZB- multiple residue-1	—	[58]
	Concentrated samples	100 mg	Residual solvents	—	—	HS-GC–MS	EI	SHRXI- 5MS	7	[60]

***BSTFA:** N,O-bis (trimethylsilyl) trifluoroacetamide; **EI:** electron ionization; **GC:** gas chromatography; **HMDS:** hexamethyldisilzane; **HS:** head space; **LLE:** liquid–liquid extraction; **MS:** mass spectrometry; **PER:** percolation; **QTOF:** quadrupole time-of-flight; **SFE:** supercritical fluid extraction; **SOX:** Soxhlet; **STU:** Soxhlet treated ultrasonication: **TMCS:** trimethylchlorosilane; **ULT:** ultrasonication; **UTS:** ultrasonication treated Soxhlet.*

Table 3.
Analytical platforms for the analysis of other minor compounds in Cannabis *samples.*

In the cultivation of *Cannabis*, pesticides are often applied to repel or eliminate unwanted pests. Despite the fact that this activity is required from an agricultural point of view, it may also affect human health due to the remain of pesticide residues in *Cannabis* and its derived products. For this reason, much attention has been focused on analyzing these compounds to guarantee the safety and quality of the final pharmaceutical drug. Different analytical methods have been applied to this end, being GC–MS one of the most common encountered techniques [21]. However, one of the main limitations associated with *Cannabis* pesticide analysis is the low-level concentration of these pollutants in the sample in comparison to cannabinoid and terpene content. For this reason, different methods have been reported to remove potential interferences and enable the analysis of residual pesticides, as LLE, SPE, SPME and quick, easy, cheap, effective, rugged, and safe (QuEChERS) [58]. Conversely, in the last years, some researchers have evaluated the potential of *Cannabis* EO as botanical insecticide. The toxicity of these products has been tested with different insect pests like aphids and houseflies, among others [51, 59]. In this way, the exploitation of hemp by-products represents an opportunity of creating a circular economy.

In the analysis of products derived from *Cannabis*, which is generally manufactured from extracts obtained with organic solvents, there are not many publications dealing with residual solvents, despite the high health risk associated with these contaminants. Probably, this is linked to the relatively recent legalization of the medical use of *Cannabis*. As a matter of fact, Raber et al. have reported the use of HS-GC–MS for the qualitative analysis of these impurities in cannabinoid concentrates [60] but it would seem plausible to think that the scarcity of recent scientific publications derives from the existence of official directions for the quantification of such residues, such as national and international alimentary codices and pharmacopeias.

5. Conclusions

Gas chromatography is a well-established tool in the *Cannabis* sector. Owing to the extraordinary health benefits of the bioactive compounds of *Cannabis sativa* L., reliable methods to quantify them are required for quality assurance. Additionally, cannabinoid and terpenoid profiles provide useful information to distinguish between *Cannabis* cultivars. Furthermore, this technique is utilized in the fight against the illegal recreational use of *Cannabis*, detecting Δ^9-THC and their metabolites in different biological samples and identifying new synthetic cannabinoids. For this reason, several analytical platforms have been developed using different modalities, such as (fast)-GC-FID, GC–MS, GC–MS/MS, GC-TOF/MS or GC-IR. Among them, GC–MS is the most utilized technique probably due to their accessibility as well as the quality of the results. Additionally, the features of this modality may be easily modified changing the ionization source, which might enhance the sensitivity and selectivity of the measurements. Thus, several ionization sources like EI, PCI, NCI and AP have been utilized in the analysis of *Cannabis*. On the other hand, although GC-TOF/MS is a more sophisticated technique, which is not as easily accessible as GC–MS, some approaches employing this modality have been already published allowing the untargeted analysis of *Cannabis* metabolites. Regarding the complex chemical composition of *Cannabis* plant, different strategies have been performed avoiding matrix effect and solving simultaneously low concentration, and chromatographic resolution problems in some cases. In addition, different extraction techniques have been utilized to determine *Cannabis* metabolites in different matrixes. Microextraction techniques like SPME, SBSE or MEPS have been applied, which involve an advantage concerning other extraction modalities since generally solvent and time consuming is minimized. In this sense, there are an extensive range of possibilities to overcome the different problems that may emerge in the analysis of *Cannabis*. Conversely, although sometimes GC approaches are non-preferred in order to avoid time-consuming derivatization steps, this process allows the analysis of instable compounds. This is the case of acidic cannabinoids, which generally are subjected to silylation processes to avoid decarboxylation in the injection port of the GC. Additionally, derivatization procedures improve the chromatographic resolution as well as the peak shape of analytes, enhancing analytical features like sensitivity and reproducibility.

Conflict of interest

The authors declare no conflict of interest.

Author details

María Teresa García-Valverde*, Verónica Sánchez de Medina, Verónica Codesido, Jesús Hidalgo-García and Carlos Ferreiro-Vera
Phytoplant Research S.L., The Science and Technology Park of Córdoba, Rabanales 21, Córdoba, Spain

*Address all correspondence to: mt.garcia@phytoplant.es

References

[1] I. J. Flores-Sanchez and A. C. Ramos-Valdivia, *A review from patents inspired by the genus Cannabis*, vol. 16, no. 4. Springer Netherlands, 2017.

[2] F. Bakro, M. Jedryczka, K. Wielgusz, B. Sgorbini, R. Inchingolo, and V. Cardenia, "Simultaneous determination of terpenes and cannabidiol in hemp (*Cannabis sativa* L.) by fast gas chromatography with flame ionization detection," *J. Sep. Sci.*, vol. 43, pp. 2817–2826, 2020, doi: 10.1002/jssc.201900822.

[3] A. Leghissa, Z. L. Hildenbrand, F. W. Foss, and K. A. Schug, "Determination of cannabinoids from a surrogate hops matrix using multiple reaction monitoring gas chromatography with triple quadrupole mass spectrometry," *J. Sep. Sci.*, vol. 41, no. 2, pp. 459–468, 2018, doi: 10.1002/jssc.201700946.

[4] D. Jin, S. Jin, Y. Yu, C. Lee, and J. Chen, "Classification of Cannabis Cultivars Marketed in Canada for Medical Purposes by Quantification of Cannabinoids and Terpenes Using HPLC-DAD and GC-MS," *J. Anal. Bioanal. Tech*, vol. 08, no. 01, p. 2, 2017, doi: 10.4172/2155-9872.1000349.

[5] F. Franchina, L. Dubois, and J.-F. Focant, "In-Depth Cannabis Multiclass Metabolite Profiling Using Sorptive Extraction and Multidimensional Gas Chromatography with Low- and High-Resolution Mass Spectrometry," *Anal. Chem.*, vol. 92, pp. 10512–10520, 2020, doi: 10.1021/acs.analchem.0c01301.

[6] V. Cardenia, *T. gallina* Toschi, S. Scappini, R. C. Rubino, and M. T. Rodriguez-Estrada, "Development and validation of a Fast gas chromatography/mass spectrometry method for the determination of cannabinoids in *Cannabis sativa* L," *J. Food Drug Anal.*, vol. 26, no. 4, pp. 1283–1292, 2018, doi: https://doi.org/10.1016/j.jfda.2018.06.001.

[7] G. Mercieca, S. Odoardi, S. Mestria, M. Cassar, and S. Strano-Rossi, "Application of ultrasound-assisted liquid–liquid microextraction coupled with gas chromatography and mass spectrometry for the rapid determination of synthetic cannabinoids and metabolites in biological samples," *J. Sep. Sci.*, 2020, doi: 10.1002/jssc.202000181.

[8] B. Fodor and I. Molnár-Perl, "The role of derivatization techniques in the analysis of plant cannabinoids by gas chromatography mass spectrometry," *TrAC Trends Anal. Chem.*, vol. 95, pp. 149–158, 2017, doi: https://doi.org/10.1016/j.trac.2017.07.022.

[9] T. Béres *et al.*, "Intralaboratory comparison of analytical methods for quantification of major phytocannabinoids," *Anal. Bioanal. Chem.*, pp. 3069–3079, 2019, doi: 10.1007/s00216-019-01760-y.

[10] N. A. dos Santos *et al.*, "Analysis of isomeric cannabinoid standards and cannabis products by UPLC-ESI-TWIM-MS: a comparison with GC-MS and GC× GC-QMS," *J. Braz. Chem. Soc.*, vol. 30, no. 1, pp. 60–70, 2019.

[11] A. Leghissa, J. Smuts, C. Qiu, Z. L. Hildenbrand, and K. A. Schug, "Detection of cannabinoids and cannabinoid metabolites using gas chromatography with vacuum ultraviolet spectroscopy," *Sep. Sci. PLUS*, vol. 1, no. 1, pp. 37–42, Jan. 2018, doi: 10.1002/sscp.201700005.

[12] V. L. Alves, J. L. Gonçalves, J. Aguiar, H. M. Teixeira, and J. S. Câmara, "The synthetic cannabinoids phenomenon: from structure to toxicological properties. A review," *Crit. Rev. Toxicol.*, vol. 50, no. 5, pp. 359–382,

2020, doi: 10.1080/10408444.2020.1762539.

[13] C. Lanz, J. Mattsson, F. Stickel, J.-F. Dufour, and R. Brenneisen, "Determination of the Endocannabinoids Anandamide and 2-Arachidonoyl Glycerol with Gas Chromatography-Mass Spectrometry: Analytical and Preanalytical Challenges and Pitfalls," *Med. Cannabis Cannabinoids*, vol. 1, no. 1, pp. 9–18, 2018, doi: 10.1159/000489032.

[14] V. R. Acquaro Junior, G. A. Goméz-Ríos, M. Tascon, M. E. Costa Queiroz, and J. Pawliszyn, "Analysis of endocannabinoids in plasma samples by biocompatible solid-phase microextraction devices coupled to mass spectrometry," *Anal. Chim. Acta*, vol. 1091, pp. 135–145, 2019, doi: 10.1016/j.aca.2019.09.002.

[15] T. Obata, Y. Sakurai, Y. Kase, Y. Tanifuji, and T. Horiguchi, "Simultaneous determination of endocannabinoids (arachidonylethanolamide and 2-arachidonylglycerol) and isoprostane (8-epiprostaglandin F2α) by gas chromatography-mass spectrometry-selected ion monitoring for medical samples," *J. Chromatogr. B Anal. Technol. Biomed. Life Sci.*, vol. 792, no. 1, pp. 131–140, 2003, doi: 10.1016/S1570-0232(03)00311-8.

[16] S. Hardison, S. T. Weintraub, and A. Giuffrida, "Quantification of endocannabinoids in rat biological samples by GC/MS: Technical and theoretical considerations," *Prostaglandins Other Lipid Mediat.*, vol. 81, no. 3–4, pp. 106–112, 2006, doi: 10.1016/j.prostaglandins.2006.08.002.

[17] G. G. Muccioli and N. Stella, "An optimized GC-MS method detects nanomolar amounts of anandamide in mouse brain," *Anal. Biochem.*, vol. 373, no. 2, pp. 220–228, 2008, doi: 10.1016/j.ab.2007.09.030.

[18] A. A. Kayacelebi *et al.*, "Cross-validated stable-isotope dilution GC–MS and LC–MS/MS assays for monoacylglycerol lipase (MAGL) activity by measuring arachidonic acid released from the endocannabinoid 2-arachidonoyl glycerol," *J. Chromatogr. B Anal. Technol. Biomed. Life Sci.*, vol. 1047, pp. 151–159, 2017, doi: 10.1016/j.jchromb.2016.08.004.

[19] J. Keereetaweep and K. D. Chapman, "Lipidomic analysis of endocannabinoid signaling: Targeted metabolite identification and quantification," *Neural Plast.*, vol. 2016, 2016, doi: 10.1155/2016/2426398.

[20] C. Marchioni, I. D. de Souza, V. R. Acquaro, J. A. de Souza Crippa, V. Tumas, and M. E. C. Queiroz, "Recent advances in LC-MS/MS methods to determine endocannabinoids in biological samples: Application in neurodegenerative diseases," *Anal. Chim. Acta*, vol. 1044, pp. 12–28, 2018, doi: 10.1016/j.aca.2018.06.016.

[21] S. N. Atapattu and K. R. D. Johnson, "Pesticide analysis in cannabis products," *J. Chromatogr. A*, vol. 1612, p. 460656, 2020, doi: 10.1016/j.chroma.2019.460656.

[22] J. Basas-Jaumandreu and F. X. C. De Las Heras, "GC-MS Metabolite Profile and Identification of Unusual Homologous Cannabinoids in High Potency *Cannabis sativa*," *Planta Med.*, vol. 86, no. 5, pp. 338–347, 2020, doi: 10.1055/a-1110-1045.

[23] C. Citti *et al.*, "A novel phytocannabinoid isolated from *Cannabis sativa* L. with an in vivo cannabimimetic activity higher than Δ9-tetrahydrocannabinol: Δ9-Tetrahydrocannabiphorol," *Sci. Rep.*, vol. 9, no. 1, pp. 1–13, 2019, doi: 10.1038/s41598-019-56785-1.

[24] R. Franco *et al.*, "Pharmacological potential of varinic-, minor-, and acidic

phytocannabinoids," *Pharmacol. Res.*, vol. 158, 2020, doi: 10.1016/j.phrs.2020.104801.

[25] R. Iseppi *et al.*, "Chemical characterization and evaluation of the antibacterial activity of essential oils from fibre-type *Cannabis sativa* L. (Hemp)," *Molecules*, vol. 24, no. 12, p. 2302, 2019.

[26] N. Fernández, L. J. Carreras, R. A. Larcher, A. S. Ridolfi, and P. N. Quiroga, "Quantification of Cannabinoids in Cannabis Oil Using GC/MS: Method Development, Validation, and Application to Commercially Available Preparations in Argentina," *Planta Medica Int. Open*, vol. 07, no. 02, pp. e81–e87, 2020, doi: 10.1055/a-1155-6613.

[27] A. Casiraghi *et al.*, "Extraction Method and Analysis of Cannabinoids in Cannabis Olive Oil Preparations," *Planta Med.*, vol. 84, no. 4, pp. 242–249, 2018, doi: 10.1055/s-0043-123074.

[28] K. Leiman, L. Colomo, S. Armenta, M. de la Guardia, and F. A. Esteve-Turrillas, "Fast extraction of cannabinoids in marijuana samples by using hard-cap espresso machines," *Talanta*, vol. 190, pp. 321–326, 2018, doi: https://doi.org/10.1016/j.talanta.2018.08.009.

[29] Y. Nuapia, H. Tutu, L. Chimuka, and E. Cukrowska, "Selective Extraction of Cannabinoid Compounds from Cannabis Seed Using Pressurized Hot Water Extraction," *Molecules*, vol. 25, p. 1335, 2020.

[30] R. Pavlovic *et al.*, "Quality traits of 'cannabidiol oils': cannabinoids content, terpene fingerprint and oxidation stability of European commercially available preparations," *Molecules*, vol. 23, no. 5, p. 1230, 2018.

[31] I. Angeli, S. Casati, A. Ravelli, M. Minoli, and M. Orioli, "A novel single-step GC–MS/MS method for cannabinoids and 11-OH-THC metabolite analysis in hair," *J. Pharm. Biomed. Anal.*, vol. 155, pp. 1–6, 2018, doi: 10.1016/j.jpba.2018.03.031.

[32] A. Gasse, H. Pfeiffer, H. Köhler, and J. Schürenkamp, "Development and validation of a solid-phase extraction method using anion exchange sorbent for the analysis of cannabinoids in plasma and serum by gas chromatography-mass spectrometry," *Int. J. Legal Med.*, vol. 130, no. 4, pp. 967–974, 2016, doi: 10.1007/s00414-016-1368-6.

[33] G. de O. Silveira, S. Loddi, C. D. R. de Oliveira, A. D. Zucoloto, L. V. G. Fruchtengarten, and M. Yonamine, "Headspace solid-phase microextraction and gas chromatography−mass spectrometry for determination of cannabinoids in human breast milk," *Forensic Toxicol.*, vol. 35, no. 1, pp. 125–132, 2017, doi: 10.1007/s11419-016-0346-5.

[34] T. Kieliba, O. Lerch, H. Andresen-Streichert, M. A. Rothschild, and J. Beike, "Simultaneous quantification of THC-COOH, OH-THC, and further cannabinoids in human hair by gas chromatography–tandem mass spectrometry with electron ionization applying automated sample preparation," *Drug Test. Anal.*, vol. 11, no. 2, pp. 267–278, Feb. 2019, doi: 10.1002/dta.2490.

[35] R. Gottardo, D. Sorio, M. Ballotari, and F. Tagliaro, "First application of atmospheric-pressure chemical ionization gas chromatography tandem mass spectrometry to the determination of cannabinoids in serum," *J. Chromatogr. A*, vol. 1591, pp. 147–154, 2019, doi: https://doi.org/10.1016/j.chroma.2019.01.041.

[36] E. A. Ibrahim *et al.*, "Determination of Acid and Neutral Cannabinoids in Extracts of Different Strains of *Cannabis sativa* Using GC-FID," *Planta Med.*, vol.

84, no. 4, pp. 250–259, 2018, doi: 10.1055/s-0043-124088.

[37] T. Franklin, L. Perry, W. Shih, and J. Yu, "Detection of phytocannabinoids from buccal swabs by headspace solid phase microextraction – gas chromatography / mass spectrometry," *Anal. Methods*, vol. 10, pp. 942–946, 2018, doi: 10.1039/C7AY02996A.

[38] M. S. Castaneto, A. Wohlfarth, N. A. Desrosiers, R. L. Hartman, D. A. Gorelick, and M. A. Huestis, "Synthetic cannabinoids pharmacokinetics and detection methods in biological matrices," *Drug Metab. Rev.*, vol. 47, no. 2, pp. 124–174, Apr. 2015, doi: 10.3109/03602532.2015.1029635.

[39] L. N. Al-Eitan, A. S. Asa'ad, A. K. H. Battah, and H. A. Aljamal, "Application of Gas Chromatography-Mass Spectrometry for the Identification and Quantitation of Three Common Synthetic Cannabinoids in Seized Materials from the Jordanian Market," *ACS Omega*, vol. 5, no. 8, pp. 4172–4180, 2020, doi: 10.1021/acsomega.9b03881.

[40] A. Namera, M. Kawamura, A. Nakamoto, T. Saito, and M. Nagao, "Comprehensive review of the detection methods for synthetic cannabinoids and cathinones," *Forensic Toxicol.*, vol. 33, no. 2, pp. 175–194, 2015, doi: 10.1007/s11419-015-0270-0.

[41] M. Akutsu, K. Sugie, and K. Saito, "Analysis of 62 synthetic cannabinoids by gas chromatography–mass spectrometry with photoionization," *Forensic Toxicol.*, vol. 35, no. 1, pp. 94–103, 2017.

[42] E. Smolianitski-Fabian, E. Cohen, M. Dronova, A. Voloshenko-Rossin, and O. Lev, "Discrimination between closely related synthetic cannabinoids by GC–Cold–EI–MS," *Drug Test. Anal.*, vol. 10, no. 3, pp. 474–487, Mar. 2018, doi: 10.1002/dta.2247.

[43] S. Gwak, L. E. Arroyo-Mora, and J. R. Almirall, "Qualitative analysis of seized synthetic cannabinoids and synthetic cathinones by gas chromatography triple quadrupole tandem mass spectrometry," *Drug Test. Anal.*, vol. 7, no. 2, pp. 121–130, Feb. 2015, doi: 10.1002/dta.1667.

[44] J. DeRuiter, F. Smith, Y. Abiedalla, L. Neel, and C. R. Clark, "GC-MS and GC-IR analysis of regioisomeric cannabinoids related to 1-(5-fluoropentyl)-3-(1-naphthoyl)-indole," *Forensic Chem.*, vol. 10, pp. 48–57, 2018, doi: https://doi.org/10.1016/j.forc.2018.07.005.

[45] R. Umebachi *et al.*, "Detection of synthetic cannabinoids using GC-EI-MS, positive GC-CI-MS, and negative GC-CI-MS," *Int. J. Legal Med.*, vol. 131, no. 1, pp. 143–152, 2017, doi: 10.1007/s00414-016-1428-y.

[46] K. Saito *et al.*, "Confirmation of synthetic cannabinoids in herb and blood by HS-SPME-GC/MS," *Forensic Chem.*, vol. 13, p. 100156, 2019, doi: https://doi.org/10.1016/j.forc.2019.100156.

[47] A. Sorribes-Soriano, J. Verdeguer, A. Pastor, S. Armenta, and F. A. Esteve-Turrillas, "Determination of Third Generation Synthetic Cannabinoids in Oral Fluids," *J. Anal. Toxicol.*, Jul. 2020, doi: 10.1093/jat/bkaa091.

[48] E. A. Ibrahim *et al.*, "Analysis of terpenes in *Cannabis sativa* L. Using GC/MS: method development, validation, and application," *Planta Med.*, vol. 85, no. 05, pp. 431–438, 2019.

[49] J. T. Fischedick, "Identification of Terpenoid Chemotypes Among High (−)- trans -Δ 9 - Tetrahydrocannabinol-Producing *Cannabis sativa* L. Cultivars," *Cannabis Cannabinoid Res.*, vol. 2, no. 1, pp. 34–47, 2017, doi: 10.1089/can.2016.0040.

[50] M. M. Delgado-Povedano, C. Sánchez-Carnerero Callado, F. Priego-Capote, and C. Ferreiro-Vera, "Untargeted characterization of extracts from *Cannabis sativa* L. cultivars by gas and liquid chromatography coupled to mass spectrometry in high resolution mode," *Talanta*, vol. 208, no. September 2019, p. 120384, 2019, doi: 10.1016/j.talanta.2019.120384.

[51] R. Ascrizzi, M. Iannone, G. Cinque, A. Marianelli, L. Pistelli, and G. Flamini, "'Hemping' the drinks: Aromatizing alcoholic beverages with a blend of *Cannabis sativa* L. flowers," *Food Chem.*, vol. 325, no. November 2019, 2020, doi: 10.1016/j.foodchem.2020.126909.

[52] L. Calvi *et al.*, "Comprehensive quality evaluation of medical *Cannabis sativa* L. inflorescence and macerated oils based on HS-SPME coupled to GC–MS and LC-HRMS (q-exactive orbitrap®) approach," *J. Pharm. Biomed. Anal.*, vol. 150, pp. 208–219, 2018, doi: 10.1016/j.jpba.2017.11.073.

[53] A. Shapira, P. Berman, K. Futoran, O. Guberman, and D. Meiri, "Tandem Mass Spectrometric Quantification of 93 Terpenoids in Cannabis Using Static Headspace Injections," *Anal. Chem.*, vol. 91, no. 17, pp. 11425–11432, 2019, doi: 10.1021/acs.analchem.9b02844.

[54] T. D. Nguyen, S. Riordan-Short, T. T. T. Dang, R. O'Brien, and M. Noestheden, "Quantitation of Select Terpenes/Terpenoids and Nicotine Using Gas Chromatography-Mass Spectrometry with High-Temperature Headspace Sampling," *ACS Omega*, vol. 5, no. 10, pp. 5565–5573, 2020, doi: 10.1021/acsomega.0c00384.

[55] M. Ternelli *et al.*, "Innovative methods for the preparation of medical Cannabis oils with a high content of both cannabinoids and terpenes," *J. Pharm. Biomed. Anal.*, vol. 186, p. 113296, 2020, doi: 10.1016/j.jpba.2020.113296.

[56] V. Devi and S. Khanam, "Comparative study of different extraction processes for hemp (*Cannabis sativa*) seed oil considering physical, chemical and industrial-scale economic aspects," *J. Clean. Prod.*, vol. 207, pp. 645–657, 2019, doi: 10.1016/j.jclepro.2018.10.036.

[57] T. Väisänen, P. Kilpeläinen, V. Kitunen, R. Lappalainen, and L. Tomppo, "Effect of steam treatment on the chemical composition of hemp (*Cannabis sativa* L.) and identification of the extracted carbohydrates and other compounds," *Ind. Crops Prod.*, vol. 131, no. February, pp. 224–233, 2019, doi: 10.1016/j.indcrop.2019.01.055.

[58] C. B. Craven, N. Wawryk, P. Jiang, Z. Liu, and X. Li, "Pesticides and trace elements in cannabis : Analytical and environmental challenges and opportunities," *J. Environ. Sci.*, vol. 85, pp. 82–93, 2019, doi: 10.1016/j.jes.2019.04.028.

[59] G. Benelli *et al.*, "The essential oil from industrial hemp (*Cannabis sativa* L.) by-products as an effective tool for insect pest management in organic crops," *Ind. Crops Prod.*, vol. 122, no. February, pp. 308–315, 2018, doi: 10.1016/j.indcrop.2018.05.032.

[60] J. C. Raber, S. Elzinga, and C. Kaplan, "Understanding dabs : contamination concerns of cannabis concentrates and cannabinoid transfer during the act of dabbing," vol. 40, no. 6, pp. 797–803, 2015.

Chapter 6

Temperature-Dependent Linear Solvation Energy Relationship for the Determination of Gas-Liquid Partition Coefficients of Organic Compounds in Ionic Liquids

Amel Ayad, Fabrice Mutelet and Amina Negadi

Abstract

In this work, a new group contribution method was used for calculating gas-to-ionic liquid partition coefficients (log K_L) of molecular solutes in ILs with a temperature-dependent linear solvation energy relationship. About 36 group parameters are used to correlate 14,762 log K_L data points of organic compounds in ionic liquids. The experimental log K_L data have been collected from the published literature for different solutes in ionic liquids at different temperatures within the range of 293.15–396.35 K. The calculated log K_L data showed a satisfactory agreement with experimental data with an average absolute relative deviation (AARD) of 6.39%.

Keywords: inverse gas chromatography, solvation model, thermodynamic properties, ionic liquids, polarity

1. Introduction

Gas chromatography is a widely used technique for the characterization of complex systems in chemical industries but also for the determination of physico-chemical properties of solute in the stationary phase. This approach is also known as inverse gas chromatography (IGC) is used to quantify the solute-stationary interactions or the polarity of the last one. At the end of the twentieth century, numerous solvation models were proposed to represent retention data from chromatography. The most well-known models are called Linear Free Energy Relationship (LFER) or Linear Solvation Energy Relationship (LSER). The most recent representation of the LSER model proposed by Abraham [1–4] is given by Eq. (1).

$$\log SP = c + eE + sS + aA + bB + lL \tag{1}$$

Where SP is a solvation parameter related with the free energy change such as gas–liquid partition coefficient, specific retention volume, or adjusted retention

IntechOpen

time at a given temperature. The capital letters represent the solutes properties and the lowercase letters the complementary properties of the ionic liquids. The solute descriptors are the excess molar refraction E, dipolarity/polarizability S, hydrogen bond acidity basicity, A and B, respectively, and the gas–liquid partition coefficient on n-hexadecane at 298 K, L. The solute descriptors may be determined using inverse gas chromatography or estimated using a group contribution method. A databank of descriptors for about 4000 compounds may be found in the literature [5–7]. The coefficients c, e, s, a, b, and l are not simply fitting coefficients, but they reflect complementary properties of the solvent phase. These coefficients are determined by multiple linear regression of Eq. (1). This model was strongly applied to characterize complex fluids.

A large databank of experimental LSER parameters for organic compounds can be found in the literature. Platts et al. proposed to determine these parameters using group contribution methods [5, 6].

In order to improve the predictive applicability of the Abraham model for ionic liquids, Sprunger et al. [8–11] have proposed a method for predicting log K_L for solute dissolved in ILs by splitting each of the six equation coefficients as a cation plus anion sum contribution. In their paper, the gas–liquid partition of solutes in ILs is represented by:

$$\begin{aligned} \log K_L = {} & (c_{KL,\ cation} + c_{KL,\ anion}) + (e_{KL,\ cation} + e_{KL,\ anion}).\mathbf{E} \\ & +(s_{KL,\ cation} + s_{KL,\ anion}).\mathbf{S} + (a_{KL,\ cation} + a_{KL,\ anion}).\mathbf{A} \\ & +(b_{KL,\ cation} + b_{KL,anion}).\mathbf{B} + (l_{KL,\ cation} + l_{KL,\ anion}).\mathbf{L} \end{aligned} \tag{2}$$

Other approaches reported in the literature to calculate log K_L are based on group contribution methods (GC). This method is useful because we need only to know the structure of the material (IL). Revelli et al. [12] proposed to splits the cation with its alkyl chains into 21 different contributions to estimate log K_L for other IL whose cation does not define in the correlation but their groups appear in this method. A new approach was developed by Mutelet et al. [13] who proposed temperature-dependent GC-LSER (TDGC-LSER) for correlating log K_L data measured of various solutes in ILs at different temperatures. A reliable prediction was obtained with a standard deviation of 0.13 log units and a coefficient of determination R^2 of 0.995. To evaluate the predictive capacity of this model, Mutelet et al. have been used data from five ILs not included in the regression. The deviations vary from 0.09 log units and 0.27 log units depending on the number of data for the groups defined by the IL. The TDGC-LSER model is given by the following equation:

$$\begin{aligned} \log K_L = {} & -2.84418 + \sum_i^{21} n_i c_i + \sum_i^{21} n_i e_i \mathbf{E} + \sum_i^{21} n_i s_i \mathbf{S} + \sum_i^{21} n_i a_i \mathbf{A} + \sum_i^{21} n_i b_i \mathbf{B} \\ & + \sum_i^{21} n_i l_i \mathbf{L}/T \end{aligned} \tag{3}$$

In this chapter, we propose to extend the model temperature-dependent GC-LSER (TDGC-LSER) based on the group contribution method to correlate and analysis of log K_L values for different solutes in ILs as a function of the temperature from 293.15 to 396.35 K. A new decomposition but also new groups are proposed in this approach. The database includes the alkyl-based ILs as well as functionalized ILs (task-specific ILs) such as alcohols and ethers. The largest number of ILs used in this database was composed of imidazolium cation-based ILs (73 ILs). Three new cations were included, choline, quinolinium, and octanium, each cation occurs only once in the database as well as Sulphonium-Based ILs.

2. Calculation of gas-liquid partition coefficients from inverse gas chromatography data

Partition coefficients K_L maybe then calculated from the activity coefficients at infinite dilution, $\gamma_{1,2}^{\infty}$, using Eq. (4):

$$K_L = \frac{RT}{\gamma_{1,2}^{\infty} P_1^0 V_{solvent}} \tag{4}$$

3. Data sets and methodology

The objective of this study is to extend the potential applicability of the TDGC-LSER model for the prediction of log K_L to different ILs include new functional groups. The ILs used in this study and their abbreviations are listed in **Table 1**.

No.	Abbreviation	Ionic liquid Name	Reference
Imidazolium-Based ILs			
1	$[MMIM]^+$ $[MeSO_4]^-$	1-metyl-3-methylimidazolium methylsulfate	[14]
2	$[BMIM]^+$ $[MeSO_4]^-$	1-butyl-3-methylimidazolium methylsulfate	[15]
3	$[EMIM]^+$ $[F_3AC]^-$	1-ethyl-3-methylimidazolium trifluoroacetate	[16]
4	$[HMIM]^+$ $[F_3AC]^-$	1-hexyl-3-methylimidazolium trifluoroacetate	[17]
5	$[EMIM]^+$ $[Tf_2N]^-$	1-Ethyl-3-methylimidazolium bis(trifluoromethylsulfonyl) imidate	[18]
6	$[EMIM]^+$ $[Tf_2N]^-$	1-Ethyl-3-methylimidazolium bis(trifluoromethylsulfonyl) Amide	[18]
7	$[EMIM]^+$ $[Tf_2N]^-$	1-Ethyl-3-methylimidazolium bis(trifluoromethylsulfonyl) imide	[19]
8	$[EMIM]^+$ $[Tf_2N]^-$	1-Ethyl-3-methylimidazolium bis(trifluoromethylsulfonyl) imidate	[18]
9	$[EMIM]^+$ $[Tf_2N]^-$	1-Ethyl-3-methylimidazolium bis(trifluoromethylsulfonyl) Amide	[18]
10	$[MMIM]^+$ $[Tf_2N]^-$	1-methyl-3-methylimidazolium bis(trifluoromethylsulfonyl) imide	[19]
11	$[BMIM]^+$ $[Tf_2N]^-$	1-butyl-3-methylimidazolium bis(trifluoromethylsulfonyl) imide	[20]
12	$[BMIM]^+$ $[Tf_2N]^-$	1-butyl-3-methylimidazolium bis(trifluoromethylsulfonyl) imide	[19]
13	$[BMIM]^+$ $[Tf_2N]^-$	1-butyl-3-methylimidazolium bis(trifluoromethylsulfonyl) imide	[21]
14	$[BMIM]^+$ $[Tf_2N]^-$	1-butyl-3-methylimidazolium bis(trifluoromethylsulfonyl) imide	[22]
15	$[MEIM]^+$ $[Tf_2N]^-$	1-methyl-3-ethylimidazolium bis(trifluoromethylsulfonyl) amide	[23]
16	$[M_2EIM]^+$ $[Tf_2N]^-$	1,2-dimethyl-3-ethylimidazolium bis (trifluoromethylsulfonyl) amide	[23]
17	$[HMIM]^+$ $[Tf_2N]^-$	1-hexyl-3-methylimidazolium bis(trifluoromethylsulfonyl) imide	[20]

No.	Abbreviation	Ionic liquid Name	Reference
18	$[HMIM]^+$ $[Tf_2N]^-$	1-hexyl-3-methylimidazolium bis(trifluoromethylsulfonyl) imide	[24]
19	$[HMIM]^+$ $[Tf_2N]^-$	1-hexyl-3-methylimidazolium bis(trifluoromethylsulfonyl) imide	[25]
20	$[OMIM]^+$ $[Tf_2N]^-$	1-octyl-3-methylimidazolium bis(trifluoromethylsulfonyl) imide	[25]
21	$[OMIM]^+$ $[Tf_2N]^-$	1-octyl-3-methylimidazolium bis(trifluoromethylsulfonyl) imide	[22]
22	$[(CH_2)_4SO_3HMIM]^+$ $[Tf_2N]^-$	1-(4-sulfobutyl)-3-methylimidazolium bis (trifluoromethanesulfonyl)imide	[26]
23	$[EtOHMIM]^+$ $[Tf_2N]^-$	1-ethanol-3-methylimidazolium bis (trifluoromethylsulfonyl)imide	[27]
24	$[MeoeMIM]^+$ $[Tf_2N]^-$	1-(methylethylether)-3-methylimidazolium bis (trifluoromethylsulfonyl)imide	[27]
25	$[(MeO)_2IM]^+$ $[Tf_2N]^-$	1,3-dimethoxyimidazolium bis(trifluoromethylsulfonyl) imide	[27]
26	$[(CH_2)_4SO_3HMIm]^+$ $[TFO]^-$	1-(4-sulfobutyl)-3-methylimidazolium trifluoromethanesulfonate	[26]
27	$[(CH_2)_4SO_3HMIm]^+$ $[HSO_4]^-$	1-(4-sulfobutyl)-3-methylimidazolium hydrogen sulfate	[26]
28	$[BMIM]^+$ $[TDI]^-$	1-butyl-3-methylimidazolium 4,5-dicyano-2 (trifluoromethyl)imidazolide	[28]
29	$[C_2OHMIM]^+$ $[FAP]^-$	1-(2-hydroxyethyl)-3-methylimidazolium tris (pentafluoroethyl)trifluorophosphate	[29]
30	$[C_2OHMIM]^+$ $[FAP]^-$	1-(2-hydroxyethyl)-3-methylimidazolium tris (pentafluoroethyl)trifluorophosphate	[30]
31	$[EMIM]^+$ $[FAP]^-$	1-ethyl-3-methylimidazolium tris(pentafluoroethyl) trifluorophosphate	[31]
32	$[BMIM]^+$ $[N(CN)_3]^-$	1-butyl-3-methylimidazolium tricyanomethanide	[32]
33	$[EMIM]^+$ $[BF_4]^-$	1-ethyl-3-methylimidazolium tetrafluoroborate	[33]
34	$[EMIM]^+$ $[BF_4]^-$	1-ethyl-3-methylimidazolium tetrafluoroborate	[34]
35	$[BMIM]^+$ $[BF_4]^-$	1-butyl-3-methylimidazolium tetrafluoroborate	[20]
36	$[BMIM]^+$ $[BF_4]^-$	1-butyl-3-methylimidazolium tetrafluoroborate	[33]
37	$[BMIM]^+$ $[BF_4]^-$	1-butyl-3-methylimidazolium tetrafluoroborate	[35]
38	$[BMIM]^+$ $[BF_4]^-$	1-butyl-3-methylimidazolium tetrafluoroborate	[21]
39	$[HMIM]^+$ $[BF_4]^-$	1-hexyl-3-methylimidazolium tetrafluoroborate	[33]
40	$[HMIM]^+$ $[BF_4]^-$	1-hexyl-3-methylimidazolium tetrafluoroborate	[36]
41	$[OMIM][BF_4]^-$	1-octyl-3-methylimidazolium tetrafluoroborate	[33]
42	$[OMIM][BF_4]^-$	1-octyl-3-methylimidazolium tetrafluoroborate	[21]
43	$[C_2OHMIM]^+$ $[BF_4]^-$	1-(2-hydroxyethyl)-3-methylimidazolium tetrafluoroborate	[37]
44	$[PM_2IM]^+$ $[BF_4]^-$	1-propyl-2,3-dimethylimidazolium tetrafluoroborate	[38]
45	$[PM_2IM]^+$ $[BF_4]^-$	1-propyl-2,3-dimethylimidazolium tetrafluoroborate	[39]
46	$[EMIM]^+$ $[(MeO)(H)PO_2]^-$	1-ethyl-3-methylimidazolium methylphosphonate	[40]
47	$[M_2MIM]^+$ $[(MeO)(H)PO_2]^-$	1.3-dimethylimidazolium methylphosphonate	[40]

No.	Abbreviation	Ionic liquid Name	Reference
48	$[EMIM]^+$ $[EtSO_4]^-$	1-ethyl-3-methylimidazolium ethylsulfate	[41]
49	$[EMIM]^+$ $[EtSO_4]^-$	1-ethyl-3-methylimidazolium ethylsulfate	[19]
50	$[BMIM]^+$ $[OcSO_4]^-$	1-butyl-3-methylimidazolium octylsulfate	[20]
51	$[BMIM]^+$ $[OcSO_4]^-$	1-butyl-3-methylimidazolium octylsulfate	[42]
52	$[EMIM]^+$ $[SCN]^-$	1-ethyl-3-methyl-imidazolium thiocyanate	[43]
53	$[EMIM]^+$ $[SCN]^-$	1-ethyl-3-methyl-imidazolium thiocyanate	[22]
54	$[BMIM]^+$ $[SCN]^-$	1-butyl-3-methylimidazolium thiocyanate	[44]
55	$[BMIM]^+$ $[SCN]^-$	1-butyl-3-methylimidazolium thiocyanate	[22]
56	$[HMIM]^+$ $[SCN]^-$	1-hexyl-3-methylimidazolium thiocyanate	[45]
57	$[MMIM]^+$ $[CH_3OC_2H_4SO_4]^-$	1-methyl-3-methylimidazolium methoxyethylsulfate	[14]
58	$[MMIM]^+$ $[(CH_3)_2PO_4]^-$	1-methyl-3-methylimidazolium dimethylphosphate	[14]
59	$[BMIM]^+$ $[PF_6]^-$	1-butyl-3-methylimidazolium hexafluorophosphate	[46]
60	$[HMIM]^+$ $[PF_6]^-$	1-hexyl-3-methylimidazolium hexafluorophosphate	[47]
61	$[HMIM]^+$ $[PF_6]^-$	1-hexyl-3-methylimidazolium hexafluorophosphate	[47]
62	$[MOIM]^+$ $[PF_6]^-$	1-methyl-3-octylimidazolium hexafluorophosphate	[48]
63	$[EMIM]^+$ $[CF_3SO_3]^-$	1-ethyl-3-methylimidazolium trifluoromethanesulfonate	[49]
64	$[BMIM]^+$ $[CF_3SO_3]^-$	1-butyl-3-methylimidazolium trifluoromethanesulfonate	[22]
65	$[HMIM]^+$ $[CF_3SO_3]^-$	1-hexyl-3-methylimidazolium trifluoromethanesulfonate	[50]
66	$[HMIM]^+$ $[CF_3SO_3]^-$	1-hexyl-3-methylimidazolium trifluoromethanesulfonate	[22]
67	$[BMIM]^+$ $[NO_3]^-$	1-Butyl-3-methylimidazolium nitrate	[51]
68	$[OMIM]^+$ $[NO_3]^-$	1-octyl-3-methylimidazolium nitrate	[52]
Pyridinium-Based ILs			
69	$[B4MPY]^+$ $[N(CN)_2]^-$	1-butyl-4-methylpyridinium dicyanamide	[53]
70	$[BMPY]^+$ $[TDI]^-$	1-Butyl-3-methylpyridinium 4,5-dicyano-2-(trifluoromethyl)imidazolide	[28]
71	$[BMPY]^+$ $[N(CN)_3]^-$	1-butyl-4-methylpyridinium tricyanomethanide	[32]
72	$[NEPY]^+$ $[Tf_2N]^-$	N-ethylpyridinium bis(trifluoromethylsulfonyl)imide	[14]
73	$[4MBPY]^+$ $[BF_4]^-$	4-methyl-N-butylpyridinium tetrafluoroborate	[54]
74	$[4MBPY]^+$ $[BF_4]^-$	4-methyl-N-butylpyridinium tetrafluoroborate	[55]
75	$[4MBPY]^+$ $[BF_4]^-$	4-methyl-N-butylpyridinium tetrafluoroborate	[46]
76	$[BMPY]^+$ $[Tf_2N]^-$	1-butyl-3-methylpyridinium bis(trifluoromethylsulfonyl) imide	[56]
77	$[1,3BMPY]^+$ $[CF_3SO_3]^-$	1-butyl-3-methylpyridinium trifluoromethanesulfonate	[57]
78	$[C_2PY]^+$ $[Tf_2N]^-$	2-alkylpyridinium bis(trifluoromethylsulfonyl)imide	[22]
79	$[C_4PY]^+$ $[Tf_2N]^-$	4-alkylpyridinium bis(trifluoromethylsulfonyl)imide	[22]
80	$[C_5PY]^+$ $[Tf_2N]^-$	5-alkylpyridinium bis(trifluoromethylsulfonyl)imide	[22]
Pyrrolidinium-Based ILs			
81	$[BMPYR]^+$ $[N(CN)_3]^-$	1-butyl-1-methylpyrrolidinium tricyanomethanide	[58]

No.	Abbreviation	Ionic liquid Name	Reference
82	$[MeoeMPYR]^+$ $[FAP]^-$	1-(2-methoxyethyl)-1-methylpyrrolidinium tris (pentafluoroethyl)trifluorophosphate	[59]
83	$[MeoeMPYR]^+$ $[FAP]^-$	1-(2-methoxyethyl)-1-methylpyrrolidinium tris (pentafluoroethyl)trifluorophosphate	[29]
84	$[PrMPYR]^+$ $[Tf_2N]^-$	1-propyl-1-methylpyrrolidinium bis (trifluoromethylsulfonyl)imide	[60]
85	$[BMPYR]^+$ $[Tf_2N]^-$	1-butyl-1-methylpyrrolidinium bis(trifluoromethylsulfonyl) imide	[25]
86	$[BMPYR]^+$ $[Tf_2N]^-$	1-butyl-1-methylpyrrolidinium bis(trifluoromethylsulfonyl) imide	[60]
87	$[PeMPYR]^+$ $[Tf_2N]^-$	1-pentyl-1-methylpyrrolidinium Bis (trifluoromethylsulfonyl)imide	[60]
88	$[HMPYR]^+$ $[Tf_2N]^-$	1-hexyl-1-methylpyrrolidinium bis(trifluoromethylsulfonyl) imide	[61]
89	$[OMPYR]^+$ $[Tf_2N]^-$	1-octyl-1-methylpyrrolidinium bis(trifluoromethylsulfonyl) imide	[61]
90	$[BMPYR]^+$ $[CF_3SO_3]^-$	1-butyl-1-methylpyrrolidinium trifluoromethanesulfonate	[62]
Piperidinium-Based ILs			
91	$[C_5C_1PIP]^+$ $[Tf_2N]^-$	1-5-alkyl-1-methylpiperidinium bis(trifluoromethylsulfonyl) imide	[63]
92	$[C_6C_1PIP]^+$ $[Tf_2N]^-$	1-6-alkyl-1-methylpiperidinium bis (trifluoromethylsulfonyl)imide	[63]
93	$[MeoeMPIP]^+$ $[FAP]^-$	1-(2-methoxyethyl)-1-methylpiperidinium trifluorotris (perfluoroethyl)phosphate	[64]
94	$[PMPIP]^+$ $[Tf_2N]^-$	1-propyl-1-methylpiperidiniumbis(trifluoromethylsulfonyl) imide	[65]
Phosphonium-Based ILs			
95	$[H_3TdP]^+$ $[L\text{-}Lact]^-$	trihexyl(tetradecyl)phosphonium L-lactate	[66]
96	$[H_3TdP]^+$ $[+CS]^-$	trihexyl(tetradecyl)phosphonium (1S)-(+)-10-camphorsulfonate	[66]
97	$[H_3TdP]^+$ $[Tf_2N]^-$	Trihexyl(tetradecyl)phosphonium bis (trifluoromethylsulfonyl)imide	[67]
98	$[H_3TdP]^+$ $[Tf_2N]^-$	Trihexyl(tetradecyl)phosphonium bis (trifluoromethylsulfonyl)imide	[25]
Morpholinium-Based ILs			
99	$[BMMOR]^+$ $[N(CN)_3]^-$	1-butyl-1-methylmorpholinium tricyanomethanide	[68]
100	$[N\text{-}C_3OHMMOR]^+$ $[Tf_2N]^-$	4-(3-hydroxypropyl)-4-methylmorpholinium bis (trifluoromethylsulfonyl)amide	[69]
Sulphonium-Based Ils			
101	$[Et_3S]^+$ $[Tf_2N]^-$	triethylsulphonium bis(trifluoromethylsulfonyl)imide	[70]
Ammonium-Based Ils			
102	$[O_3MAM]^+$ $[Tf_2N]^-$	trioctylmethylammonium bis(trifluoromethylsulfonyl)imide	[71]
103	$[O_3MAM]^+$ $[Tf_2N]^-$	trioctylmethylammonium bis(trifluoromethylsulfonyl)imide	[71]
104	[M3BAM] + [NTf2]−	trimethyl-butylammonium bis(trifluoromethylsulfonyl) imide	[72]

No.	Abbreviation	Ionic liquid Name	Reference
Choline-Based Ils			
105	$[Cho]^+$ $[Tf_2N]^-$	choline bis(trifluoromethylsulfonyl)imide	[73]
Quinolinium-Based Ils			
106	$[HiQuin]^+$ $[SCN]^-$	N-hexylisoquinolinium thiocyanate	[74]
Octanium-Based Ils			
107	$[HDABCO]^+$ $[Tf_2N]^-$	1-hexyl-1,4-diaza[2.2.2]bicyclooctanium bis (trifluoromethylsulfonyl)imide	[75]

Table 1.
Names and abbreviations of ionic liquids.

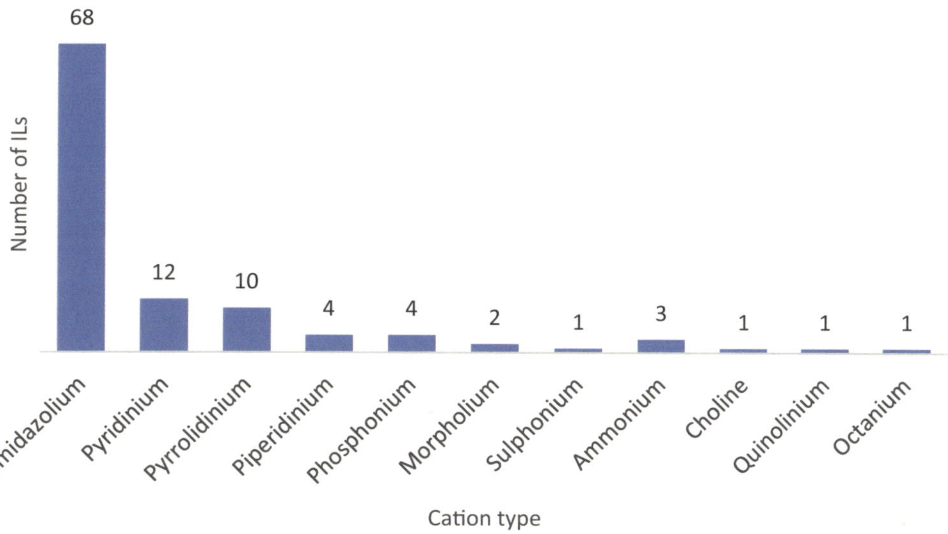

Figure 1.
Number of cation-based ILs used for the prediction of log K_L.

The families of ILs represent 11 cation types are imidazolium, pyridinium, pyrrolidinium, piperidinium, phosphonium, morpholium, sulphonium, ammonium, choline, quinolinium, octanium. From **Figure 1**, we can see ILs with respect to their cations used for the correlation of log K_L. The largest number of ILs used in this database was composed of imidazolium cation-based ILs. Three new cations were included, choline, quinolinium, and octanium. For the anions, 20 different structures are used, for example tris(pentafluoroethyl)trifluorophosphate, nitrate, bis(trifluoromethylsulfonyl)imide, hexafluorophosphate, tetrafluoroborate, and sulfates.

The collection of organic solutes includes alkanes, cycloalkanes, alkenes, alkynes, aromatics, alcohols, ethers, aldehydes, ketones, chloroalkanes cyanoalkanes, thiophene, pyridine, water, and other solutes. The values of the five solute descriptors for different organic compounds considered in this work can be found in the literature [1–6]. The solute descriptors used in the calculations were taken from earlier studies on ILs and cover the range from: $E = -0.020$ to $E = 0.851$, $S = 0.000$ to $S = 0.950$, $A = 0.000$ to $A = 0.82$, $B = 0.000$ to $B = 0.79$, $L = 0.26$ to $L = 6.705$. A databank containing 107 ILs corresponding to 14,762 experimental log K_L were used for the development of the model. The 36 groups used for the representation of ILs are given in **Table 2**.

Definition	Group	Definition	Group
CH_3 from alkyl chain	CH_3	Trifluorotris(perfluoroethyl)phosphate	$[FAP]^-$
CH_2 from alkyl chain	CH_2	Bis(trifluoromethylsulfonyl)imide	$[Tf_2N]^-$
-O- in alkyl chain	O	Hexafluorophosphate	$[PF_6]^-$
Hydroxyl in alkyl	OH	Tetrafluoroborate	$[BF_4]^-$
Sulfonyl hydroxide	SO_3H	Methylsulfate	$[MeSO_4]^-$
N ammonium's cation	$[N]^+$	Ethylsulfate	$[EtSO_4]^-$
S sulfonium's cation	$[S]^+$	Octylsulfate	$[OcSO_4]^-$
1,4-diaza [2] bicyclooctanium	$[DABCO]^+$	Thiocyanate	$[SCN]^-$
Imidazolium cation	$[IM]^+$	Trifluoromethylsulfonate	$[CF_3SO_3]^-$
Pyridinium cation	$[PY]^+$	Trifluoroacetate	$[F_3AC]^-$
Pyrrolidinium cation	$[PYR]^+$	Methoxyethylsulfate	$[CH_3OC_2H_4SO_4]^-$
Piperidinium cation	$[PIP]^+$	4,5-Dicyano-2-(trifluoromethyl) imidazolium	$[TDI]^-$
Phosphonium cation	$[P]^+$	Methylphosphonate	$[(MeO)(H)PO_2]^-$
Morpholium cation	$[MOR]^+$	Hydrogen sulfate	$[HSO_4]^-$
Quinolinium cation	$[Quin]^+$	Dimethylphosphate	$[(CH_3)_2PO_4]^-$
Choline cation	$[Cho]^+$	L-lactate	$[_L\text{-Lact}]^-$
Dicyanamide	$[N(CN)_2]^-$	(1S)-(+)-10-Camphorsulfonate	$[+CS]^-$
Tricyanomethanide	$[N(CN)_3]^-$	Nitrate	$[NO_3]^-$

Table 2.
Description of the 36 groups used for the estimation of log K_L*.*

4. Results and discussion

In the present study, the database contains 14,762 experimental data for about 107 ILs at the temperature range of 293.15–396.35 K, the main objective of this study is to extend the predictive applicability of the TDGC-LSER model to a new cation and anion based ILs, that have not been reported previously, as the ILs having cyano-based anions (thiocyanate $[SCN]^-$; dicyanamide, $[N(CN)_2]^-$; tricyano-methanide, $[C(CN)_3]^-$ and perfluorinated anions (e.g. trifluorotris(perfluoroethyl) phosphate, [FAP]). The model adopted in this work is capable of representing log K_L over a wide range of temperatures. The quality of regression analysis was determined by calculating the average absolute relative deviation (AARD) defined as follows:

$$AARD(logK_L) = \frac{1}{N}\sum_{i=1}^{N}\left|\frac{logK_L^{cal} - logK_L^{exp}}{logK_L^{exp}}\right| \times 100 \quad (5)$$

The TDGC-LSER model may represent the partition coefficient of solutes in ILs using the Eq. 6:

$$\log K_L = -3.1729 + \sum_i^{36} n_i c_i + \sum_i^{36} n_i e_i \mathbf{E} + \sum_i^{36} n_i s_i \mathbf{S} + \sum_i^{36} n_i a_i \mathbf{A} + \sum_i^{36} n_i b_i \mathbf{B} + \sum_i^{36} n_i l_i \mathbf{L}/T \quad (6)$$

Where n_i is the number of group i present in the ionic liquid. Values of c_i, e_i, s_i, a_i, b_i, and l_i and their standard errors in parentheses are given in **Table 3**.

The experimental log K_L data for all ILs were reproduced using Eq. (6) with average absolute deviation (AARD) at the level of 6.39%. The model developed is statistically good, and describes the experimental log K_L to within a standard deviation of 0.134 log units, squared correlation coefficient R^2 = 96.9%, and a Fisher's F statistic F = 2106.87.

Group	c_i	e_i	s_i	a_i	b_i	l_i
CH_3	−82.195 (8.697)	67.71 (13.14)	1.38 (15.33)	133.26 (16.50)	5.69 (14.63)	4.18 (2.594)
CH_2	11.093 (1.523)	−25.428 (2.319)	−15.933 (2.836)	−14.017 (2.827)	−3.663 (3.065)	8.57 (0.4577)
O	−18.818 (6.703)	−9.581 (8.681)	21.984 (8.577)	−30.93 (11.21)	21.494 (9.081)	−13.017 (1.732)
OH	−193 (12.76)	98.11 (17.90)	39.42 (20.01)	168.51 (22.02)	293.86 (19.65)	−17.394 (3.687)
$[N]^+$	2316.75 (64.96)	353.55 (90.42)	−297.80 (112.6)	−387.6 (120.5)	98.1 (123.9)	−263.57 (17.99)
$[S]^+$	1892.89 (47.02)	−69.03 (61.36)	71.74 (73.65)	−358.17 (80.73)	190.98 (67.59)	−59.51 (11.93)
SO_3H	−401.01 (27.18)	211.14 (25.90)	29.56 (29.25)	963.61 (46.37)	646.47 (35.68)	−0.109 (5.620)
DABCO	1804.97 (41.71)	116.87 (41.93)	−89.01 (46.46)	−141.72 (70.03)	232.09 (53.11)	−60.643 (9.825)
$[IM]^+$	1929.34 (39.51)	−29.8 (39.59)	−8.73 (44.00)	−267.56 (65.70)	180.46 (47.66)	−63.12 (9.213)
$[PY]^+$	1899.81 (39.00)	8.09 (39.56)	−20.39 (44.43)	−254.68 (65.27)	189.68 (47.58)	−59.425 (9.001)
$[PYR]^+$	1891.63 (40.38)	−5.33 (40.98)	16.51 (45.52)	−222.69 (67.02)	117.9 (49.23)	−50.992 (9.509)
$[PIP]^+$	1896.55 (40.90)	6.56 (42.01)	30.75 (46.89)	−231.35 (67.60)	65.98 (50.61)	−46.801 (9.668)
$[P]^+$	1942.67 (72.12)	410.56 (98.99)	48.1 (116.4)	−402.1 (128.0)	191.7 (122.5)	−243.72 (20.34)
$[MOR]^+$	1834.41 (41.57)	44.92 (42.28)	107.86 (46.83)	−119.24 (68.33)	88.14 (50.55)	−72.741 (9.943)
$[Quin]^+$	1849.71 (40.76)	25.7 (42.69)	33.16 (48.08)	−145.85 (66.40)	33.22 (49.85)	−46.729 (9.584)
$[Cho]^+$	−183.29 (35.99)	−522.76 (58.38)	333.14 (78.77)	−30.32 (75.04)	−49.82 (93.81)	185.55 (10.51)
$[N(CN)_2]^-$	−970.82 (36.63)	150.74 (33.05)	810.89 (35.05)	1249.76 (61.39)	−81.18 (40.34)	197.042 (8.294)
$[N(CN)_3]^-$	−896.18 (34.95)	82.95 (28.43)	725.96 (29.59)	894.88 (56.33)	13.96 (36.19)	199.506 (7.598)
$[FAP]^-$	−830.7 (34.70)	−74.84 (29.00)	736.99 (30.24)	372.32 (56.58)	52.4 (36.71)	214.343 (7.509)
$[Tf_2N]^-$	−899.66 (33.67)	−29.2 (27.04)	734.55 (28.17)	640.99 (55.10)	−10.5 (34.81)	207.618 (7.095)

Group	c_i	e_i	s_i	a_i	b_i	l_i
$[PF_6]^-$	−977.77 (36.33)	85.23 (38.82)	758.26 (47.71)	528.15 (64.97)	125.04 (49.75)	208.42 (8.180)
$[BF_4]^-$	−958.9 (34.01)	144.71 (27.81)	733.41 (29.17)	979.13 (56.03)	−40.58 (35.85)	187.379 (7.214)
$[MeSO_4]^-$	−912.02 (38.32)	179.36 (34.98)	439.56 (38.88)	1866.13 (72.09)	112.7 (49.26)	147.143 (9.112)
$[EtSO_4]$—	−959.72 (36.50)	−12.58 (35.09)	792.73 (39.00)	1467.31 (65.24)	−194.98 (48.93)	198.684 (7.880)
$[OcSO_4]^-$	−883.39 (45.40)	−65.96 (41.05)	575.78 (42.09)	1395.39 (73.88)	−231.21 (50.54)	249.11 (11.64)
$[SCN]^-$	−1120.06 (34.61)	239.28 (32.03)	720.87 (36.59)	1388.16 (56.41)	64.13 (38.49)	210.298 (7.459)
$[CF_3SO_3]^-$	−971.54 (34.72)	84.5 (30.51)	708.98 (34.13)	1048.13 (57.45)	35.85 (39.56)	205.302 (7.504)
$[F_3AC]^-$	−948.31 (36.54)	192.18 (32.62)	516.67 (34.42)	1654.6 (60.42)	−41.21 (40.15)	201.913 (8.340)
$[CH_3OC_2H_4SO_4]^-$	−864.62 (60.40)	−117.89 (90.33)	1150.9 (142.0)	−190.7 (776.4)	−563.9 (186.0)	128.97 (19.99)
$[TDI]^-$	−900.02 (34.69)	8.82 (28.82)	719.91 (29.78)	868.4 (56.35)	−50.29 (36.22)	228.757 (7.537)
$[(MeO)(H)PO_2]^-$	−837.19 (39.44)	−110.29 (34.18)	772.28 (36.82)	1843.42 (68.33)	−103.52 (42.76)	175.47 (8.406)
$[HSO_4]^-$	−718.34 (91.93)	−438.37 (52.51)	991.64 (66.58)	−457.3 (283.4)	−1113.4 (101.0)	171.09 (16.10)
$[(CH_3)_2PO_4]^-$	−1189.85 (64.54)	327.11 (52.20)	379.11 (68.71)	1887.4 (79.33)	445.78 (81.67)	242.14 (17.61)
[L-Lact]$^-$	−935.34 (46.66)	−35.01 (45.39)	869.85 (46.65)	2176.62 (89.53)	−254.38 (59.97)	227.99 (11.50)
[+CS]$^-$	−1008.33 (44.06)	−29.46 (43.56)	822.58 (43.45)	1674.07 (74.84)	−160.92 (51.68)	228.68 (10.48)
$[NO_3]^-$	−937.17 (33.75)	215.29 (26.72)	565.62 (29.28)	1483.74 (45.69)	125.91 (29.89)	188.276 (7.003)

Table 3.
TDGC-LSER coefficients for the calculation of log K_L.

The average absolute deviation AARD (γ^∞) on the prediction of activity coefficients at infinite dilution γ^∞ for each IL considered in this study are determined by the following equation:

$$AARD(\gamma^\infty) = \frac{1}{N}\sum_{i=1}^{N}\left|\frac{\gamma_{cal}^\infty - \gamma_{exp}^\infty}{\gamma_{exp}^\infty}\right| \times 100 \tag{7}$$

The AARD values on the prediction of log K_L and γ^∞ for each IL considered in this work are reported in **Table 4**.

The values obtained for each ILs vary from 0.87 to 35.41%. **Figure 2** shows a plot of calculated log K_L from TDGC-LSER versus experimental log K_L for 107 ILs. The derived equation provides a good description of solute transfer into cyano-based anion such as $[BMIM]^+$ $[C(CN)_3]^-$, $[BMPY]^+$ $[C(CN)_3]^-$, $[BMPYR]^+$ $[C(CN)_3]^-$ for which the values of AARD are at the level of about 3% except for $[BMMOR]^+$

Ionic liquids	Log K_L (%)	$\gamma_{1,2}^{\infty}$ (%)	Ionic liquids	Log K_L (%)	$\gamma_{1,2}^{\infty}$ (%)
Imidazolium-Based ILs					
$[MMIM]^+$ $[MeSO_4]^-$	22.82	56.93	$[EMIM]^+$ $[EtSO_4]^-$	6.71	21.63
$[BMIM]^+$ $[MeSO_4]^-$	13.17	27.03		10.20	19.59
$[EMIM]^+$ $[F_3AC]^-$	5.91	15.16	$[BMIM]^+$ $[OcSO_4]^-$	6.70	33.60
$[HMIM]^+$ $[F_3AC]^-$	4.54	32.12		2.78	14.49
$[MMIM]^+$ $[Tf_2N]^-$	6.56	26.59	$[MMIM]^+$ $[CH_3OC_2H_4SO_4]^-$	13.82	11.28
$[EMIM]^+$ $[Tf_2N]^-$	13.55	98.45	$[MMIM]^+$ $[(CH_3)_2PO_4]^-$	10.73	29.15
	9.36	54.05	$[EMIM]^+$ $[CF_3SO_3]^-$	3.24	11.55
	5.38	29.87	$[BMIM]^+$ $[CF_3SO_3]^-$	3.38	11.03
	13.69	101.29	$[HMIM]^+$ $[CF_3SO_3]^-$	10.11	23.18
	10.64	64.23		6.93	21.79
$[MEIM]^+$ $[Tf_2N]^-$	6.10	17.00	$[BMIM]^+$ $[NO_3]^-$	6.31	29.82
$[M_2EIM]^+$ $[Tf_2N]^-$	6.73	24.15	$[OMIM]^+$ $[NO_3]^-$	6.12	32.88
$[BMIM]^+$ $[Tf_2N]^-$	3.90	22.78	*Pyridinium-Based ILs*		
	5.11	24.57	$[B4MPY]^+$ $[N(CN)_2]^-$	4.54	12.64
	6.99	39.58	$[BMPY]^+$ $[TDI]^-$	3.35	14.75
	3.54	27.55	$[BMPY]^+$ $[C(CN)_3]^-$	3.77	19.51
$[HMIM]^+$ $[Tf_2N]^-$	3.15	21.65	$[BMPY]^+$ $[Tf_2N]^-$	5.18	16.77
	3.17	16.04	$[4MBPY]^+$ $[BF_4]^-$	3.88	21.36
	2.90	15.23		4.82	15.22
	2.86	20.70		2.67	16.47
	0.87	5.15	$[1,3BMPY]^+$ $[CF_3SO_3]^-$	9.98	42.38
$[OMIM]^+$ $[Tf_2N]^-$	3.12	11.35	$[NEPY]^+$ $[Tf_2N]^-$	9.54	29.23
	2.62	11.09	$[C_2PY]^+$ $[Tf_2N]^-$	5.37	21.46
$[(CH_2)_4SO_3HMIm]^+$ $[Tf_2N]^-$	7.40	34.57	$[C_4PY]^+$ $[Tf_2N]^-$	9.63	32.88
$[EtOHmim]^+$ $[Tf_2N]^-$	7.12	26.81	$[C_5PY]^+$ $[Tf_2N]^-$	9.63	32.88
$[(MeO)_2IM]^+$ $[Tf_2N]^-$	8.05	21.27	*Pyrrolidinium-Based ILs*		
$[MeoeMIM]^+$ $[Tf_2N]^-$	4.70	20.21	$[BMPYR]^+$ $[C(CN)_3]^-$	3.09	11.06
$[EMIM]^+$ $[BF_4]^-$	9.64	60.61	$[MeoeMPyrr]^+$ $[FAP]^-$	4.70	18.65
	11.71	38.35		5.61	29.68
$[PM_2IM]^+$ $[BF_4]^-$	6.18	23.63	$[PrMPYR]^+$ $[Tf_2N]^-$	4.23	16.02
	35.41	35.51	$[BMPYR]^+$ $[Tf_2N]^-$	4,10	15.54
$[BMIM]^+$ $[BF_4]^-$	5.15	11.76	$[PeMPYR]^+$ $[Tf_2N]^-$	3.79	14.22
	12.10	22.42	$[HMPYR]^+$ $[Tf_2N]^-$	1.95	7.89
	4.25	21.49	$[OMPYR]^+$ $[Tf_2N]^-$	3.27	12.20
	5.32	29.16	$[BMPYR]^+$ $[CF_3SO_3]^-$	7.97	14.51
	5.55	28.39	$[BMPYR]^+$ $[Tf_2N]^-$	4.60	22.93
$[HMIM]^+$ $[BF_4]^-$	6,02	19.93	*Piperidinium-Based ILs*		
	9.63	51.73	$[PMPIP]^+$ $[Tf_2N]^-$	3,19	13.44

Ionic liquids	Log K_L (%)	$\gamma^{\infty}_{1,2}$ (%)	Ionic liquids	Log K_L (%)	$\gamma^{\infty}_{1,2}$ (%)
$[OMIM]^+ [BF_4]^-$	3.57	14.14	$[C_5C_1PiP]^+ [Tf_2N]^-$	2.42	10.89
	3.71	22.32	$[C_6C_1PIP]^+ [Tf_2N]^-$	2.47	10.12
	5.69	29.40	$[MeoeMPIP]^+ [FAP]^-$	3.74	15.66
	11.37	90.97	*Phosphonium-Based ILs*		
$[C_2OHMIM]^+ [BF_4]^-$	10.65	38,39	$[H_3TdP]^+$ [L-Lact]$^-$	3.06	16.80
$[BMIM]^+ [PF_6]^-$	2.70	18.20	$[H_3TdP]^+$ [+CS]$^-$	5.90	33.28
$[HMIM]^+ [PF_6]^-$	3.36	14.79	$[H_3TdP]^+ [Tf_2N]^-$	3.58	16.19
	3.32	14.51		4.55	23.55
$[MOIM]^+ [PF_6]^-$	2.91	12.12	*Morpholinium-Based ILs*		
$[(CH_2)_4SO_3HMIm]^+ [TFO]^-$	12.52	31.50	$[BMMOR]^+ [C(CN)_3]^-$	5.94	13.17
$[(CH_2)_4SO_3HMIm]^+ [HSO_4]^-$	8.56	36.71	$[N\text{-}C_3OHMMOR]^+ [Tf_2N]^-$	12.42	15.87
$[BMIM]^+ [TDI]^-$	3.42	13.01	*Sulphonium-Based ILs*		
$[EMIM]^+ [FAP]^-$	8.41	23.43	$[Et3S]^+ [Tf_2N]^-$	3.63	8.95
$[C_2OHMIM]^+ [FAP]^-$	8.25	54.98	*Ammonium-Based ILs*		
	4.71	21.68	$[M3BA]^+ [Tf_2N]^-$	11.08	91.05
$[EMIM]^+ [SCN]^-$	14.50	14.31	$[O3MA]^+ [Tf_2N]^-$	8.61	52.55
	9.17	19.02	$[O3MA]^+ [Tf_2N]^-$	8.72	50.13
$[BMIM]^+ [SCN]^-$	10.20	18.10	*Choline-Based ILs*		
	14.21	29.36	$[Cho]^+ [Tf_2N]^-$	4.82	13.95
$[HMIM]^+ [SCN]^-$	5.44	18.68	*Quinolinium-Based ILs*		
$[BMIM]^+ [C(CN)_3]^-$	3.14	10.86	$[HiQuin]^+ [SCN]^-$	7.93	16.65
$[EMIM]^+ [(MeO)(H)PO_2]^-$	15.89	50.54	*Octanium-Based ILs*		
$[DIMIM]^+ [(MeO)(H)PO_2]^-$	9.70	47.47	$[HDABCO]^+ [Tf_2N]^-$	4.28	18.48

Table 4.
Average absolute deviation for the ionic liquids that are used in this study.

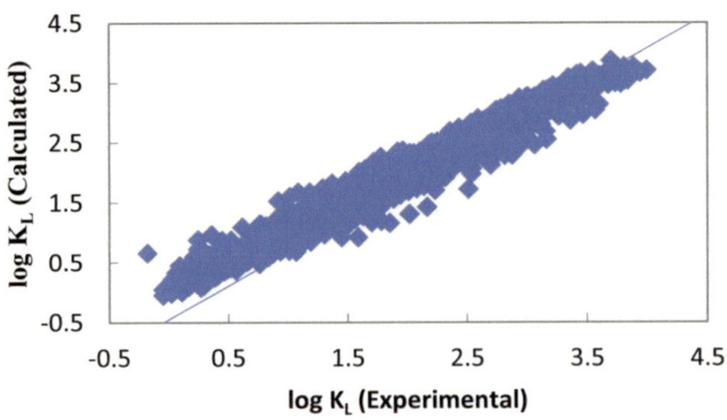

Figure 2.
Comparison of calculated log K_L values based on Eq. (6) versus experimental data.

$[C(CN)_3]^-$ with AARD of 5.94%. For the IL $[HMIM]^+ [Tf_2N]^-$ the AARD was less than 1%. The same all piperidinium cation-based ILs studied in this work have a small AARD lower than 4%.

The ILs with the highest AARD are $[MMIM]^+$ $[MeSO_4]^-$, $[PM_2IM]^+$ $[BF_4]^-$, $[BMIM]^+$ $[NO_3]^-$, $[OMIM]^+$ $[NO_3]^-$, This may be related to the quality of the experimental data or the number of experimental data.

5. Prediction of partition coefficients of organic compounds in ILs not included in the database using the TDGC-LSER model

The predictive power of TDGC-LSER was evaluated calculating log K_L of organic compounds in four ILs not included in the database: 1-butyl-3-methylimidazolium chloride $[BMIM]^+$ $[Cl]^-$ [76], 1-butyl-3-methylimidazolium dimethyl phosphate $[BMIM]^+$ $[(CH_3)_2PO_4]^-$ [76], 1-butyl-3-methylimidazolium dicyanamide $[BMIM]^+$ $[N(CN)_2]^-$ [77], 1-Dodecyl-3-methylimidzolium Bis(trifluoromethylsulfonyl)-imide $[DoMIM]^+$ $[Tf_2N]^-$ [78]. In the case of $[BMIM]^+$ $[Cl]^-$, 224 experimental log K_L values were predicted using the TDGC-LSER model. The difference between the calculated values and the experimental data is significant since it reaches 44.8%. In the case of alkanes, this difference is greater than 100% (for cyclopentane (173.24) and ethyl tert-butyl ether (120.38) at T = 358.15 K). The poor performance of the model can be explained by the limited number of experimental data containing the group $[Cl]^-$. In addition, this ionic liquid is particularly hydrophilic and can therefore contain significant amounts of water. This presence of water can have a significant impact on the physicochemical properties of this ionic liquid. For $[BMIM]^+$ $[(CH_3)_2PO_4]^-$, the AARD of log K_L is 17.84%. The results indicate that the model correctly predicts the partition coefficients of aromatics, methanol, and ethanol. The log K_L values of alkanes and ethers are underestimated. This can be explained by the low solubility of alkanes in the ILs and also the limited number of ethers in the database. The AARD observed with $[BMIM]^+$ $[N(CN)_2]^-$ for the log K_L prediction of 378 solutes is 19.78%. The TDGC-LSER model provides a good description for alkenes, alkynes, aromatic hydrocarbons (o-xylene about 4.72% at T = 318.15), alcohols, and ketones (acetone about 5.13% at 348.15 K). For most alkanes, the log K_L values are underestimated, in particular for pentane (109.32% at 338.15 K), 2.2-dimethylbutane, 2.2.4-triethylpentane and cyclopentane. $[DoMIM]^+$ $[Tf_2N]^-$ contains a long alkyl chain grafted onto the imidazolium cation. 284 experimental log K_L values were estimated using the TDGC-LSER model. The correlation provides a good description of the experimental log K_L data (AARD = 15.77%). This result is important as it shows that the model can be used to predict solute partition coefficients in ILs not taken into account in the database. The model provides a good description for polar solutes such as aromatics (benzene 4.46% at T = 318.15 K) and alcohols. The most significant differences are observed for the ethers (tert-butyl methyl ether, Diethyl ether, Di-n-propyl ether, Di-iso-propyl ether).

6. Conclusions

The TDGC-LSER model was used for the prediction of log K_L for 107 ILs over a wide range of temperatures. A new set of cation and anion functional groups have been used to define the structures of the ionic liquids. The obtained correlation allows reproduce the experimental gas-to-ionic liquid partition coefficients derived from γ^∞ with satisfactory accuracy to within 0.134 log units and AARD of 6.39%. The deviation between calculated and experimental log K_L depends strongly on the cation and anion type.

The list of groups used for the estimation of calculated log K_L in this work can be increased to improve the predictive applicability of this model to other functional groups by the additional γ^∞ data for families of a new generation of ionic liquids.

Author details

Amel Ayad[1], Fabrice Mutelet[2]* and Amina Negadi[1]

1 Université de Tlemcen, Tlemcen, Algeria

2 Ecole Nationale Supérieure des Industries Chimiques Université de Lorraine, Nancy, France

*Address all correspondence to: fabrice.mutelet@univ-lorraine.fr

References

[1] Abraham MH. Scales of solute hydrogen-bonding: Their construction and application to physicochemical and biochemical processes. Chemical Society Reviews. 1993;**22**(2):73-83. DOI: 10.1039/CS9932200073

[2] Abraham MH, Grellier PL, McGill RA. Determination of olive oil–gas and hexadecane–gas partition coefficients, and calculation of the corresponding olive oil–water and hexadecane–water partition coefficients. Journal of the Chemical Society, Perkin Transactions. 1987;**2**(6): 797-803. DOI: 10.1039/P29870000797

[3] Abraham MH, Whiting GS, Doherty RM, Shuely WJ. Hydrogen bonding. XVI. A new solute salvation parameter, π2H, from gas chromatographic data. Journal of Chromatography A. 1991;**587**(2): 213-228. DOI: 10.1016/0021-9673(91) 85158-C

[4] Abraham MH, Whiting GS, Doherty RM, Shuely WJ. Hydrogen bonding. Part 13. A new method for the characterisation of GLC stationary phases—The Laffort data set. Journal of the Chemical Society, Perkin Transactions. 1990;**2**(8):1451-1460

[5] Abraham MH, Platts JA. Hydrogen bond structural group constants. Journal of Organic Chemistry. 2001;**66**(10): 3484-3491. DOI: 10.1021/jo001765s

[6] Platts JA, Butina D, Abraham MH, Hersey A. Estimation of molecular linear free energy relation descriptors using a group contribution approach. Journal of Chemical Information and Computer Sciences. 1999;**39**(5):835-845. DOI: 10.1021/ci980339t

[7] Mutelet F, Rogalski M. Experimental determination and prediction of the gas-liquid n-hexadecane partition coefficients. Journal of Chromatography A. 2001;**923**(1–2): 153-163. DOI: 10.1016/S0021-9673(01) 00995-5

[8] Sprunger LM, Proctor A, Acree WE Jr, Abraham MH. LFER correlations for room temperature ionic liquids: Separation of equation coefficients into individual cation-specific and anion-specific contributions. Fluid Phase Equilibria. 2008;**265**(1–2):104-111. DOI: 10.1016/j.fluid.2008.01.006

[9] Sprunger L, Clark M, Acree WE Jr, Abraham MH. Characterization of room-temperature ionic liquids by the abraham model with cation-specific and anion-specific equation coefficients. Journal of Chemical Information and Modeling. 2007;**47**(3):1123-1129. DOI: 10.1021/Ci7000428

[10] Sprunger LM, Gibbs J, Proctor A, Acree WE Jr, Abraham MH, Meng Y, et al. Linear free energy relationship correlations for room temperature ionic liquids: Revised cation-specific and anion-specific equation coefficients for predictive applications covering a much larger area of chemical space. Industrial and Engineering Chemistry Research. 2009;**48**(8):4145-4154. DOI: 10.1021/ ie801898j

[11] Proctor A, Sprunger L, Acree WE Jr, Abraham MH. LFER correlations for the solubilising characterisation of room temperature ionic liquids containing trifluoromethanesulfonate and trifluoroacetate anions. Physics and Chemistry of Liquids. 2008;**46**(6): 631-642. DOI: 10.1080/0031910080 2087191

[12] Revelli A-L, Mutelet F, Jaubert J-N. Prediction of partition coefficients of organic compounds in ionic liquids: Use of a linear solvation energy relationship with parameters calculated through a group contribution method. Industrial

and Engineering Chemistry Research. 2010;**49**(8):3883-3892. DOI: 10.1021/ie901776z

[13] Mutelet F, Ortega-Villa V, Moïse J-C, Jaubert J-N, Acree WE. Prediction of partition coefficients of organic compounds in ionic liquids using a temperature-dependent linear solvation energy relationship with parameters calculated through a group contribution method. Journal of Chemical and Engineering Data. 2011;**56**(9): 3598-3606. DOI: 10.1021/je200454d

[14] Kato R, Gmehling J. Activity coefficients at infinite dilution of various solutes in the ionic liquids [MMIM]+ [CH3SO4]−, [MMIM]+ [CH3OC2H4SO4]−, [MMIM]+ [(CH3) 2PO4]−, [C5H5NC2H5]+ [(CF3SO2) 2N]− and [C5H5NH]+ [C2H5OC2H4OSO3]−. Fluid Phase Equilibria. 2004;**226**:37-44. DOI: 10.1016/j.fluid.2004.08.039

[15] Ge M-L, Deng X-M, Zhang L-H, Chen J-Y, Xiong J-M, Li W-H. Activity coefficients at infinite dilution of organic solutes in the ionic liquid 1-butyl-3-methylimidazolium methyl sulfate. The Journal of Chemical Thermodynamics. 2014;**77**:7-13. DOI: 10.1016/j.jct.2014.04.020

[16] Domańska U, Marciniak A. Activity coefficients at infinite dilution measurements for organic solutes and water in the ionic liquid 1-Ethyl-3-methylimidazolium trifluoroacetate. The Journal of Physical Chemistry. B. 2007;**111**(41):11984-11988. DOI: 10.1021/jp075079+

[17] Jiang L-K, Wang L-S, Du C-J, Wang X-Y. Activity coefficients at infinite dilution of organic solutes in 1-hexyl-3-methylimidazolium trifluoroacetate and influence of interfacial adsorption using gas–liquid chromatography. The Journal of Chemical Thermodynamics. 2014;**70**: 138-146. DOI: 10.1016/j.jct.2013.10.038

[18] Deenadayalu N, Letcher TM, Reddy P. Determination of activity coefficients at infinite dilution of polar and nonpolar solutes in the ionic liquid 1-Ethyl-3-methyl- imidazolium bis (trifluoromethylsulfonyl) Imidate using gas−liquid chromatography at the temperature 303.15 K or 318.15 K. Journal of Chemical & Engineering Data. 2005;**50**(1):105-108. DOI: 10.1021/je0498107

[19] Krummen M, Wasserscheid P, Gmehling J. Measurement of activity coefficients at infinite dilution in ionic liquids using the dilutor technique. Journal of Chemical & Engineering Data. 2002;**47**(6):1411-1417. DOI: 10.1021/je0200517

[20] Zhou Q, Wang L-S. Activity coefficients at infinite dilution of alkanes, alkenes, and alkyl benzenes in 1-Butyl-3-methylimidazolium tetrafluoroborate using gas−liquid chromatography. Journal of Chemical & Engineering Data. 2006;**51**(5): 1698-1701. DOI: 10.1021/je060142u

[21] Zhang J, Zhang Q, Qiao B, Deng Y. Solubilities of the gaseous and liquid solutes and their thermodynamics of solubilization in the novel room-temperature ionic liquids at infinite dilution by gas chromatography. Journal of Chemical & Engineering Data. 2007; **52**(6):2277-2283. DOI: 10.1021/je700297c

[22] Yan P-F, Liu Q-S, Yang M, Liu X-M, Tan Z-C, Welz-Biermann U. Activity coefficients at infinite dilution of organic solutes in N-alkylpyridinium bis (trifluoromethylsulfonyl)imide ([CnPY][NTf2], n=2,4,5) using gas–liquid chromatography. The Journal of Chemical Thermodynamics. 2010; **42**(12):1415-1422. DOI: 10.1016/j.jct.2010.06.009

[23] Heintz A, Kulikov DV, Verevkin SP. Thermodynamic properties of mixtures containing ionic liquids. 2. Activity

coefficients at infinite dilution of hydrocarbons and polar solutes in 1-Methyl-3-ethyl-imidazolium bis (trifluoromethyl-sulfonyl) amide and in 1,2-Dimethyl-3-ethyl-imidazolium bis (trifluoromethyl-sulfonyl) amide using gas–liquid chromatography. Journal of Chemical & Engineering Data. 2002; **47**(4):894-899. DOI: 10.1021/je0103115

[24] Letcher TM, Marciniak A, Marciniak M, Domańska U. Activity coefficients at infinite dilution measurements for organic solutes in the ionic liquid 1-hexyl-3-methyl-imidazolium bis(trifluoromethyl sulfonyl)-imide using g.l.c. at T=(298.15, 313.15, and 333.15) K. The Journal of Chemical Thermodynamics. 2005;**37**(12):1327-1331. DOI: 10.1016/j.jct.2005.03.014

[25] Kato R, Gmehling J. Systems with ionic liquids: Measurement of VLE and γ∞ data and prediction of their thermodynamic behavior using original UNIFAC, mod. UNIFAC(do) and COSMO-RS(Ol). The Journal of Chemical Thermodynamics. 2005;**37**(6): 603-619. DOI: 10.1016/j.jct.2005.04.010

[26] Bensaid Z, Mutelet F, Bouroukba M, Negadi A. Experimental and theoretical study of interaction between organic compounds and 1-(4-sulfobutyl)-3-methylimidazolium based ionic liquids. Fluid Phase Equilibria. 2014;**378**: 34-43. DOI: 10.1016/j.fluid.2014.06.021

[27] Revelli A-L, Mutelet F, Jaubert J-N, Garcia-Martinez M, Sprunger LM, Acree WE, et al. Study of ether-, alcohol-, or Cyano-functionalized ionic liquids using inverse gas chromatography. Journal of Chemical & Engineering Data. 2010;**55**(7): 2434-2443. DOI: 10.1021/je900838a

[28] Zawadzki M, Niedzicki L, Wieczorek W, Domańska U. Estimation of extraction properties of new imidazolide anion based ionic liquids on the basis of activity coefficient at infinite dilution measurements. Separation and Purification Technology. 2013;**118**:242-254. DOI: 10.1016/j.seppur.2013.07.006

[29] Órfão EF, Dohnal V, Blahut A. Infinite dilution activity coefficients of volatile organic compounds in two ionic liquids composed of the tris (pentafluoroethyl)trifluorophosphate ([FAP]) anion and a functionalized cation. The Journal of Chemical Thermodynamics. 2013;**65**:53-64. DOI: 10.1016/j.jct.2013.05.035

[30] Marciniak A, Wlazło M. Activity coefficients at infinite dilution and physicochemical properties for organic solutes and water in the ionic liquid 1-(2-hydroxyethyl)-3-methylimidazolium trifluorotris(perfluoroethyl)phosphate. The Journal of Chemical Thermodynamics. 2013;**64**:114-119. DOI: 10.1016/j.jct.2013.05.008

[31] Wlazło M, Marciniak A, Letcher TM. Activity coefficients at infinite dilution and physicochemical properties for organic solutes and water in the ionic liquid 1-Ethyl-3-methylimidazolium trifluorotris(perfluoroethyl)phosphate. Journal of Solution Chemistry. 2015; **44**(3):413-430. DOI: 10.1007/s10953-014-0274-0

[32] Lukoshko E, Mutelet F, Domanska U. Experimental and theoretically study of interaction between organic compounds and tricyanomethanide based ionic liquids. The Journal of Chemical Thermodynamics. 2015;**85**:49-56. DOI: 10.1016/j.jct.2014.12.027

[33] Foco GM, Bottini SB, Quezada N, de la Fuente JC, Peters CJ. Activity coefficients at infinite dilution in 1-Alkyl-3-methylimidazolium tetrafluoroborate ionic liquids. Journal of Chemical & Engineering Data. 2006; **51**(3):1088-1091. DOI: 10.1021/je050544m

[34] Ge M-L, Wang L-S, Wu J-S, Zhou Q. Activity coefficients at infinite dilution of organic solutes in 1-Ethyl-3-methylimidazolium tetrafluoroborate using gas–liquid chromatography. Journal of Chemical & Engineering Data. 2008;**53**(8):1970-1974. DOI: 10.1021/je800218g

[35] Zhou Q, Wang L-S, Wu J-S, Li M-Y. Activity coefficients at infinite dilution of polar solutes in 1-Butyl-3-methylimidazolium tetrafluoroborate using gas–liquid chromatography. Journal of Chemical & Engineering Data. 2007;**52**(1):131-134. DOI: 10.1021/je060305e

[36] Letcher TM, Soko B, Reddy P, Deenadayalu N. Determination of activity coefficients at infinite dilution of solutes in the ionic liquid 1-Hexyl-3-methylimidazolium tetrafluoroborate using gas–liquid chromatography at the temperatures 298.15 K and 323.15 K. Journal of Chemical & Engineering Data. 2003;**48**(6):1587-1590. DOI: 10.1021/je030187k

[37] Li Y, Wang L-S, Zhang Y. Activity coefficients at infinite dilution of polar solutes in 1-(2-hydroxyethyl)-3-methylimidazolium tetrafluoroborate using gas–liquid chromatography. Journal of Chemical & Engineering Data. 2010;**55**(4):1732-1734. DOI: 10.1021/je900704b

[38] Ge M-L, Wu J-S, Wang M-H, Wang L-S. Activity coefficients at infinite dilution of polar solutes in 1-Propyl-2,3-dimethylimidazolium tetrafluoroborate using gas–liquid chromatography. Journal of Chemical & Engineering Data. 2008;**53**(3):871-873. DOI: 10.1021/je700640r

[39] Wang M-H, Wu J-S, Wang L-S, Li M-Y. Activity coefficients at infinite dilution of alkanes, alkenes, and alkyl benzenes in 1-Propyl-2,3-dimethylimidazolium tetrafluoroborate using gas–liquid chromatography. Journal of Chemical & Engineering Data. 2007;**52**(4):1488-1491. DOI: 10.1021/je6005696

[40] Ayad A, Mutelet F, Abumandour E-S, Negadi A. Activity coefficients at infinite dilution of organic solutes in methylphosphonate based ionic liquids using gas-liquid chromatography. The Journal of Chemical Thermodynamics. 2015;**86**:116-122. DOI: 10.1016/j.jct.2015.02.023

[41] Sumartschenkowa IA, Verevkin SP, Vasiltsova TV, Bich E, Heintz A, Shevelyova MP, et al. Experimental study of thermodynamic properties of mixtures containing ionic liquid 1-Ethyl-3-methylimidazolium ethyl sulfate using gas-liquid chromatography and transpiration method. Journal of Chemical & Engineering Data. 2006;**51**(6):2138-2144

[42] Mutelet F, Jaubert J-N. Accurate measurements of thermodynamic properties of solutes in ionic liquids using inverse gas chromatography. Journal of Chromatography A. 2006;**1102**(1):256-267. DOI: 10.1016/j.chroma.2005.10.046

[43] Domańska U, Marciniak A. Measurements of activity coefficients at infinite dilution of aromatic and aliphatic hydrocarbons, alcohols, and water in the new ionic liquid [EMIM][SCN] using GLC. The Journal of Chemical Thermodynamics. 2008;**40**(5):860-866. DOI: 10.1016/j.jct.2008.01.004

[44] Domańska U, Laskowska M. Measurements of activity coefficients at infinite dilution of aliphatic and aromatic hydrocarbons, alcohols, thiophene, tetrahydrofuran, MTBE, and water in ionic liquid [BMIM][SCN] using GLC. The Journal of Chemical Thermodynamics. 2009;**41**(5):645-650. DOI: 10.1016/j.jct.2008.12.018

[45] Domańska U, Marciniak A, Królikowska M, Arasimowicz M. Activity coefficients at infinite dilution measurements for organic solutes and water in the ionic liquid 1-Hexyl-3-methylimidazolium thiocyanate. Journal of Chemical & Engineering Data. 2010; **55**(7):2532-2536. DOI: 10.1021/je900890u

[46] Shimoyama Y, Hirayama T, Iwai Y. Measurement of infinite dilution activity coefficients of alcohols, ketones, and aromatic hydrocarbons in 4-methyl-N-butylpyridinium tetrafluoroborate and 1-Butyl-3-methylimidazolium hexafluoro phosphate by gas–liquid chromato graphy. Journal of Chemical & Engineering Data. 2008;**53**(9): 2106-2111. DOI: 10.1021/je800246v

[47] Letcher TM, Soko B, Ramjugernath D, Deenadayalu N, Nevines A, Naicker PK. Activity coefficients at infinite dilution of organic solutes in 1-Hexyl-3-methylimidazolium hexafluorophosphate from gas–liquid chromatography. Journal of Chemical & Engineering Data. 2003;**48**(3):708-711. DOI: 10.1021/je0256481

[48] Olivier E, Letcher TM, Naidoo P, Ramjugernath D. Activity coefficients at infinite dilution of organic solutes in the ionic liquid 1-octyl-3-methylimidazolium hexafluorophosphate using gas–liquid chromatography at T=(313.15, 323.15, and 333.15) K. The Journal of Chemical Thermodynamics. 2010;**42**(5):646-650. DOI: 10.1016/j.jct.2009.12.004

[49] Olivier E, Letcher TM, Naidoo P, Ramjugernath D. Activity coefficients at infinite dilution of organic solutes in the ionic liquid 1-ethyl-3-methylimida zolium trifluoromethanesulfonate using gas–liquid chromatography at T= (313.15, 323.15, and 333.15) K. The Journal of Chemical Thermodynamics. 2010;**42**(1):78-83. DOI: 10.1016/j.jct.2009.07.010

[50] Yang X-J, Wu J-S, Ge M-L, Wang L-S, Li M-Y. Activity coefficients at infinite dilution of alkanes, alkenes, and alkyl benzenes in 1-Hexyl-3-methy limidazolium Trifluoromethane sulfonate using gas–liquid chromato graphy. Journal of Chemical & Engineering Data. 2008;**53**(5): 1220-1222. DOI: 10.1021/je800043a

[51] Feng Y-X, Wang L-S, Li Y. Activity coefficients at infinite dilution of organic solutes in 1-Butyl-3-methylimidazolium nitrate using gas–liquid chromatography. Journal of Chemical & Engineering Data. 2011; **56**(5):2730-2736. DOI: 10.1021/je200050q

[52] Duan J-D, Wang L-S, Jiang K, Wang X-X. Activity coefficients at infinite dilution of organic solutes in 1-octyl-3-methylimidazolium nitrate using gas–liquid chromatography. Fluid Phase Equilibria. 2012;**328**:1-8. DOI: 10.1016/j.fluid.2012.05.006

[53] The study of activity coefficients at infinite dilution for organic solutes and water in 1-butyl-4-methyl pyridinium dicyanamide, [B4MPy] [DCA] using GLC. The Journal of Chemical Thermodynamics. 2014; **68**:138-144. DOI: 10.1016/j.jct.2013.09.007

[54] Heintz A, Kulikov DV, Verevkin SP. Thermodynamic properties of mixtures containing ionic liquids. Activity coefficients at infinite dilution of polar solutes in 4-methyl- N-butyl-pyridinium tetrafluoroborate using gas–liquid chromatography. The Journal of Chemical Thermodynamics. 2002; **34**(8):1341-1347. DOI: 10.1006/jcht.2002.0961

[55] Heintz A, Kulikov DV, Verevkin SP. Thermodynamic properties of mixtures containing ionic liquids. 1. Activity

coefficients at infinite dilution of alkanes, alkenes, and alkylbenzenes in 4-methyl-n-butylpyridinium tetrafluoroborate using gas–liquid chromatography. Journal of Chemical & Engineering Data. 2001;**46**(6): 1526-1529. DOI: 10.1021/je0101348

[56] Domańska U, Marciniak A. Activity coefficients at infinite dilution measurements for organic solutes and water in the ionic liquid 4-methyl-N-butyl-pyridinium bis (trifluoromethylsulfonyl)-imide. The Journal of Chemical Thermodynamics. 2009;**41**(12):1350-1355. DOI: 10.1016/j.jct.2009.06.011

[57] Marciniak A, Wlazlo M. Activity coefficients at infinite dilution measurements for organic solutes and water in the ionic liquid 1-Butyl-3-methyl-pyridinium Trifluoromethanesulfonate. Journal of Chemical & Engineering Data. 2010; **55**(9):3208-3211. DOI: 10.1021/je1000582

[58] Domańska U, Lukoshko EV. Measurements of activity coefficients at infinite dilution for organic solutes and water in the ionic liquid 1-butyl-1-methylpyrrolidinium tricyanomethanide. The Journal of Chemical Thermodynamics. 2013;**66**: 144-150. DOI: 10.1016/j.jct.2013.07.004

[59] Marciniak A, Wlazło M. Activity coefficients at infinite dilution and physicochemical properties for organic solutes and water in the ionic liquid 1-(2-methoxyethyl)-1-methylpyrrolidinium trifluorotris (perfluoroethyl)phosphate. The Journal of Chemical Thermodynamics. 2013;**60**: 57-62. DOI: 10.1016/j.jct.2013.01.007

[60] Mutelet F, Hassan E-SRE, Stephens TW, Acree WE, Baker GA. Activity coefficients at infinite dilution for organic solutes dissolved in three 1-Alkyl-1-methylpyrrolidinium bis (trifluoromethylsulfonyl)imide ionic liquids bearing short linear alkyl side chains of three to five carbons. Journal of Chemical & Engineering Data. 2013; **58**(8):2210-2218. DOI: 10.1021/je4001894

[61] Nebig S, Liebert V, Gmehling J. Measurement and prediction of activity coefficients at infinite dilution (γ∞), vapor–liquid equilibria (VLE) and excess enthalpies (HE) of binary systems with 1,1-dialkyl-pyrrolidinium bis(trifluoromethylsulfonyl)imide using mod. UNIFAC (Dortmund). Fluid Phase Equilibria. 2009;**277**(1):61-67. DOI: 10.1016/j.fluid.2008.11.013

[62] Domańska U, Redhi GG, Marciniak A. Activity coefficients at infinite dilution measurements for organic solutes and water in the ionic liquid 1-butyl-1-methylpyrrolidinium trifluoromethanesulfonate using GLC. Fluid Phase Equilibria. 2009;**278**(1): 97-102. DOI: 10.1016/j.fluid.2009.01.011

[63] Paduszyński K, Domańska U. Experimental and theoretical study on infinite dilution activity coefficients of various solutes in piperidinium ionic liquids. The Journal of Chemical Thermodynamics. 2013;**60**:169-178. DOI: 10.1016/j.jct.2013.01.005

[64] Marciniak A, Wlazło M. Activity coefficients at infinite dilution and physicochemical properties for organic solutes and water in the ionic liquid 1-(2-methoxyethyl)-1-methylpiperidinium trifluorotris (perfluoroethyl)phosphate. The Journal of Chemical Thermodynamics. 2013;**57**: 197-202. DOI: 10.1016/j.jct.2012.08.016

[65] Domańska U, Paduszyński K. Measurements of activity coefficients at infinite dilution of organic solutes and water in 1-propyl-1-methylpiperidinium bis{(trifluoromethyl)sulfonyl}imide ionic liquid using g.l.c. The Journal of Chemical Thermodynamics. 2010; **42**(11):1361-1366. DOI: 10.1016/j.jct.2010.05.017

[66] Mutelet F, Alonso D, Stephens TW, Acree WE, Baker GA. Infinite dilution activity coefficients of solutes dissolved in two Trihexyl(tetradecyl) phosphonium ionic liquids. Journal of Chemical & Engineering Data. 2014; **59**(6):1877-1885. DOI: 10.1021/je500050p

[67] Tumba K, Letcher TM, Naidoo P, Ramjugernath D. Activity coefficients at infinite dilution of organic solutes in the ionic liquid trihexyltetradecyl phosphonium bis (trifluoromethyl sulfonyl) imide using gas–liquid chromatography at T=(313.15, 333.15, 353.15 and 373.15) K. The Journal of Chemical Thermodynamics. 2013;**65**: 159-167. DOI: 10.1016/j.jct.2013.05.030

[68] Domańska U, Lukoshko EV. Thermodynamics and activity coefficients at infinite dilution for organic solutes and water in the ionic liquid 1-butyl-1-methylmorpholinium tricyanomethanide. The Journal of Chemical Thermodynamics. 2014;**68**: 53-59. DOI: 10.1016/j.jct.2013.08.030

[69] Wlazło M, Marciniak A, Zawadzki M, Dudkiewicz B. Activity coefficients at infinite dilution and physicochemical properties for organic solutes and water in the ionic liquid 4-(3-hydroxypropyl)-4-methyl morpholinium bis(trifluoromethy lsulfonyl)-amide. The Journal of Chemical Thermodynamics. 2015;**86**: 154-161. DOI: 10.1016/j.jct.2015.02.024

[70] Domańska U, Marciniak A. Activity coefficients at infinite dilution measurements for organic solutes and water in the ionic liquid triethylsulphonium bis (trifluoromethylsulfonyl)imide. The Journal of Chemical Thermodynamics. 2009;**41**(6):754-758. DOI: 10.1016/j.jct.2008.12.005

[71] Gwala NV, Deenadayalu N, Tumba K, Ramjugernath D. Activity coefficients at infinite dilution for solutes in the trioctylmethylammonium bis(trifluoromethylsulfonyl)imide ionic liquid using gas–liquid chromatography. The Journal of Chemical Thermodynamics. 2010;**42**(2):256-261. DOI: 10.1016/j.jct.2009.08.012

[72] Heintz A, Vasiltsova TV, Safarov J, Bich E, Verevkin SP. Thermodynamic properties of mixtures containing ionic liquids. 9. Activity coefficients at infinite dilution of hydrocarbons, alcohols, esters, and aldehydes in trimethyl-butylammonium bis (trifluoromethylsulfonyl) imide using gas–liquid chromatography and static method. Journal of Chemical & Engineering Data. 2006;**51**(2):648-655. DOI: 10.1021/je050440b

[73] Domańska U, Papis P, Szydłowski J. Thermodynamics and activity coefficients at infinite dilution for organic solutes, water and diols in the ionic liquid choline bis (trifluoromethylsulfonyl)imide. The Journal of Chemical Thermodynamics. 2014;**77**:63-70. DOI: 10.1016/j.jct.2014.04.024

[74] Królikowska M, Karpińska M, Królikowski M. Measurements of activity coefficients at infinite dilution for organic solutes and water in N-hexylisoquinolinium thiocyanate, [HiQuin][SCN] using GLC. The Journal of Chemical Thermodynamics. 2013;**62**: 1-7. DOI: 10.1016/j.jct.2013.02.004

[75] Marcinkowski Ł, Kloskowski A, Namieśnik J. Measurement of activity coefficients at infinite dilution of organic solutes in the ionic liquid 1-hexyl-1,4-diaza[2.2.2]bicyclooctanium bis(trifluoromethylsulfonyl)imide using gas–liquid chromatography. The Journal of Chemical Thermodynamics. 2014;**71**: 84-90. DOI: 10.1016/j.jct.2013.10.026

[76] Martins MAR, Coutinho JAP, Pinho SP, Domańska U. Measurements of activity coefficients at infinite

dilution of organic solutes and water on polar imidazolium-based ionic liquids. The Journal of Chemical Thermodynamics. 2015;**91**:194-203. DOI: 10.1016/j.jct.2015.07.042

[77] Domańska U, Wlazło M, Karpińska M. Activity coefficients at infinite dilution of organic solvents and water in 1-butyl-3-methylimidazolium dicyanamide. A literature review of hexane/hex-1-ene separation. Fluid Phase Equilibria. 2016;**417**:50-61. DOI: 10.1016/j.fluid.2016.02.004

[78] Domańska U, Wlazło M. Thermodynamics and limiting activity coefficients measurements for organic solutes and water in the ionic liquid 1-dodecyl-3-methylimidzolium bis (trifluoromethylsulfonyl) imide. The Journal of Chemical Thermodynamics. 2016;**103**:76-85. DOI: 10.1016/j.jct.2016.08.008